QL 496 .W34 1996

Waldbauer, Gilbert.

Insects through the seasons.

QL 496 .W34 1996

Waldbauer, Gilbert.

Insects through the seasons.

$24.95   ORIOLE PARK BRANCH 7/94

DATE	BORROWER'S NAME	
JUN 2 4 1998		
OCT 1 6 2000		
JAN 1 9 2001		
OCT 2 9 2001		

Insects through the Seasons

Harvard University Press

Cambridge, Massachusetts

London, England ⌐ 1996

Insects through the Seasons

GILBERT WALDBAUER

Drawings by Amy Bartlett Wright
Design by Marianne Perlak

Library of Congress cataloging information follows the index

To Benjamin Aaron Waldbauer Yates, who shares my love
for the little creatures of the world, and to Joseph Brauner,
Dominick D'Ostillio, and Aretas Saunders, who first
introduced me to the beauties of nature

Contents

Illustrations

Preface

From an ecological perspective, the insects are the most successful of all the animal groups. Their known species number almost a million, and there are probably several million as yet undiscovered species. They occur almost everywhere on the land and in fresh water, and they have survived on earth for hundreds of millions of years. In this book I attempt to take the measure of their success and to consider some of the evolutionary adaptations that have made it possible for them to flourish—the mechanisms through which they obtain food, find mates, escape enemies, and meet the other challenges to surviving and reproducing in complex and often unforgiving environments.

Scores of insect species appear on the following pages, but there is a leading character, a protagonist that we follow through the seasons and that serves as a basis for comparison. This protagonist, the handsome cecropia moth, is a favorite of mine, the object of much of my research for well over a decade. It is also a favorite of amateur naturalists and is one of the few insects that the general public views with equanimity—sometimes even with pleasure. Thousands of grade-school children know this moth because enterprising teachers collect cecropia cocoons in the winter. Their students are instructed and delighted by the large and colorful moths that emerge from the cocoons in the spring.

Plants, birds, and animals other than insects appear on these pages frequently. Like all living things, insects are best viewed in the light of their ecological context, the plants and other animals with which they associate. The living organisms in any ecosystem—a woodland, a pond, or even a cultivated field—form an integrated community of interdependent organisms. Living within its community, an animal may survive; separated from its community, it will surely perish. Trying to understand an insect taken out of its ecological milieu is as futile as trying to see significance in a single word lifted from its context on this page.

A great modern biologist, G. Evelyn Hutchison of Yale University, said that the history of life on earth is an evolutionary drama played on an ecological stage. Some of that drama can be inferred from the fossil record, and bits of it can be glimpsed by comparing the various ways in which different insects, or insects and other animals, cope with similar problems of survival and reproduction. As you read on, you will find many such comparisons— even comparisons between the ways of insects and the ways of people.

It has been impossible for me to write about insects without considering how they affect people. A few insects, probably less than 2 percent of the species in the world, make life difficult for us by transmitting diseases, eating our crops, or destroying our stores of food. But as a group, they are essential to our well-being, because they are indispensable components of virtually all of the ecosystems upon which we depend for the food and other organic products without which we could not survive. Even the survival of city dwellers who have never seen a farm ultimately depends upon insects. Without insects as pollinators or scavengers, for example, most of these ecosystems would disappear and be replaced by ones that are far less hospitable to humans.

In the first century A.D., the Roman encyclopedist Gaius Plinius Secundus, also known as Pliny the Elder, assured his readers that in the mountains north of India there are ants that mine gold and that are as large as wolves. There are numerous myths about insects, some even more fanciful than Pliny's tale. But the true story of the insects is even more interesting and marvelous.

Insects are endlessly fascinating—wonderful in their amazing variety and marvelous in the perfection of the evolutionary adaptations that have made them the dominant animal group on the earth. When I wake up in the morning to get ready for work, I often think about how fortunate I am to be an entomologist. I spend my days reading about insects, lecturing about them, and investigating their lives in the laboratory or in woodlands and fields. Nothing could be more pleasing to me. It is as if I were paid to do my hobby. I hope that I can communicate some of my enthusiasm for insects, and show you why I find studying them to be so appealing and absorbing.

First Things

In a dense shrub at the base of a birch tree, a large silken cocoon hung safely hidden throughout the autumn and winter. The caterpillar that had spun the cocoon fed on the birch but descended to the safety of the dense bush to spin its winter resting place. Nestled in the cocoon is the winter stage of cecropia, the pupa. Until a few days ago the flame of life in the pupa was just an infinitesimal spark. At that time the pupa was in its period of winter rest, a quiet state called diapause, the insects' version of hibernation. But now environmental events and activity within the body of the pupa have fanned the spark into a tiny flame. The pupa is awake. The wonderful transformation called metamorphosis is proceeding.

This cecropia cocoon could be located almost anywhere in the southern portions of eastern Canada or in the eastern half of the United States except in its southernmost areas. But the timing of the events in cecropia's life, as I present them, will be most accurate in a broad band that extends westward from southern New England and the mid-Atlantic states to Iowa and Missouri.

The season is early spring, the first few days of April—just three weeks after the spring equinox. Cecropia is not the only insect that has awakened by this time. After their long winter confinement, honey bees, among the few temperate zone insects that do not diapause, are out foraging for pollen from early spring blossoms. Mourning cloak butterflies spent the winter in hollow trees and other sheltered nooks but can now be seen flitting about in the woods. To the north, early-season flies and moths, which seek sweet, energy-rich meals, can be found drowned in sap buckets in a grove of maples, known as a sugar bush in New England. Far to the south, monarch butterflies are well into the migration from Mexico that will bring some of them and the descendants of others back to the United States and southern Canada in time to lay their eggs on the leaves of milkweeds in late spring. Plants that will be pollinated by early-season insects have put forth their blossoms. The spring beauties are a carpet of light pink in the woodlands; little brown bees and a variety of flies seek them out for their nectar. Dandelions dot suburban lawns. In wet places, skunk cabbage flowers, clublike masses of tightly clumped blossoms covered by a squat, mottled, purplish hood, have pushed up through the soft earth. But the leaves are still small and will remain furled until later in the spring. Like a warm-blooded animal, the clump of skunk cabbage blossoms can raise its internal temperature as much as 63° F above the temperature of the surrounding air. This very unplantlike ability probably helps it attract and detain the insects that pollinate its blossoms. Those charming little tree frogs, the spring peepers, are whistling in chorus from woodland pools where they have gathered to reproduce. Later in the season they will disperse to trees and shrubs and use their sticky tongues to capture small insects. The earliest migrant birds have returned from the south to feed on those insects that are now abundant. A phoebe darts out from its perch to snatch a fly from the air. Towhees and migrant fox sparrows scratch among last year's fallen leaves as they search for seeds and for beetles and other ground-dwelling insects. Over a pond, tree swallows swoop back and forth as they pursue midges and other flying insects. Male red-winged blackbirds have arrived to stake out their territories in cattail marshes and along weedy roadsides. A little later in the spring, when insects are more abundant, the females will arrive and take their places in the nesting harems of the males.

Let us return to our cecropia cocoon in the dense bush beneath the birch. Within the cocoon lies the pupa, the life stage between the caterpillar (larva) and the moth. It is sometimes referred to as a resting stage. There is some truth in this. The pupae of most insects—certainly the pupae of cecropia and the other moths and butterflies—are not even capable of moving about. They can make only squirming movements because their six legs and four wings are immovably fused to the surface of their bodies. The pupa of cecropia looks like a dark brown, unshelled pecan on whose surface the legs, wings, and antennae are carved in bas relief. But to think of the pupa as only a resting stage is misleading. During the winter a diapausing pupa is certainly physically at rest. But after diapause has ended in the spring, the inside of its body is a very busy place indeed. During its resting period, the pupa is filled with what appears to be an amorphous and viscous yellow mass. This mass once formed the legs, eyes, and other organs of the sluggish and wormlike caterpillar. But most of these structures were broken down when the caterpillar metamorphosed to become the pupa. Little or no structure is visible inside the resting pupa, but don't be deceived—the apparently amorphous mass already contains a sketchy framework of the organs of the adult. Once the period of diapause is over, metamorphosis will once again proceed. The sketchy structures in the pupa will be completed, the viscous mass reorganized to form the definitive legs, wings, and other organs of the active and fast-flying moth.

If we remove a cecropia pupa from its cocoon in spring, we can actually catch glimpses of the metamorphosis as it progresses within the translucent skin. Gently brushing the pupal skin with alcohol makes it easier to see what is going on inside. During the first few days, very subtle changes in the structure of the developing legs of the adult are dimly visible, but only with the magnification of a low-power microscope. At about the midpoint of the metamorphosis, the claws at the ends of the legs can be seen through a magnifying glass. Shortly thereafter, the featherlike antennae are visible to the naked eye through the skin of the pupa. Finally, just before the adult has finished its development, the color pattern of its wings is very obvious. Not long after this, a matter of perhaps hours or a day, the fully formed moth pushes its way out of the pupal skin and then out of the cocoon.

At this time of year, late March and April, metamorphosis

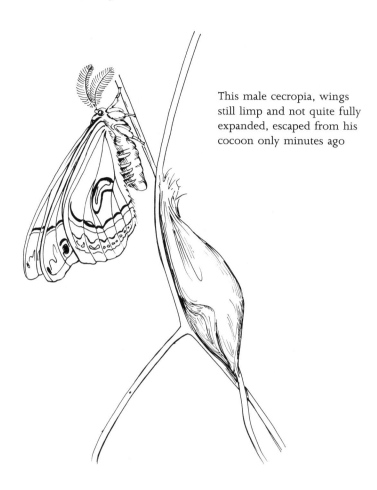

This male cecropia, wings still limp and not quite fully expanded, escaped from his cocoon only minutes ago

proceeds slowly because air temperatures are low in early spring and because cecropia, like most of the other insects, has no internal mechanism for controlling its own body temperature: it is a cold-blooded animal like a fish or a lizard. Some cold-blooded animals have behavioral mechanisms that give limited control of body temperature. Early on cool mornings, butterflies may sit on the tops of plants with their wings spread wide so as to expose them to the warmth of the sun. Many desert insects are active only during the cool of the night; during the day they dig deep into the soil to escape the hot, deadly rays of the sun. A few insects can control their body temperature internally; they are essentially "warm-blooded." Among them are bumble bees and certain dung beetles.

By late May, almost two months later, metamorphosis will be complete and the adult cecropia will be ready to shed the skin of the pupa and to make its way out of the cocoon to face the world. The season is now sufficiently advanced to permit these moths to get on with their lives. At night the air temperature is usually high enough so that the nocturnal and cold-blooded moths can fly. The males will fly to seek a mate. The females will, after they mate, fly in search of the trees and woody shrubs on whose leaves they lay their eggs. By late May, these leaves are fully grown and ready to serve as food for the young caterpillars that will hatch from the eggs. But hostile as well as benevolent elements arrive with the advancing spring. Bats, nocturnal like cecropia, have come out of hibernation and are hungry for insects. Late-arriving migrant birds such as the tanagers and the orioles, which may find and devour a resting moth in daylight, have begun their reproductive cycle and will soon require an abundant supply of insects, probably including cecropia caterpillars, to feed their hungry nestlings.

In late May, in this case early in the last week of the month, our moth emerges from its cocoon. Cecropia moths emerge in the morning, on the average at about ten o'clock. The newly emerged moth is a shrunken and wretched version of its later resplendent self. The body is wet with the fluids that lubricated its passage from the pupal skin and out into the world through a valve at the head-end of the cocoon. The wings are small, as yet unexpanded bundles of soft and delicate tissue. At this stage, they would be irreparably damaged by the touch of a finger. The moth crawls a few inches up the stem to which the cocoon is attached and then stops moving. The wings become firm and hard after blood pressure has expanded them to their full size and shape, and although it is now capable of flying, the moth simply sits on its hidden perch in the shrub. Depending upon its sex, it will

remain essentially motionless for at least the rest of the daylight hours.

⟍ If our moth is a male, the next activity occurs just after dusk, about eleven hours later in the day. At this time, the male will fly away from the dense shrub. No one really knows everything that the males do after dusk, but they probably fly for as much as an hour and then hide in a shrub or tree somewhere far from their cocoons for the rest of the night. The male will become active again about an hour before dawn, when it will fly off in search of a female. If our moth is a female, she will remain inactive throughout that day and the following night. About an hour before dawn, at about the same time that the male becomes active for the second time, she will do something for the first time since she expanded her wings some seventeen hours earlier. You must look closely to see this activity. She just protrudes a small gland from the tip of her abdomen. The gland emits a scent, a pheromone that will attract males. The scent drifts downwind, invisible but nevertheless forming a long expanding plume like smoke drifting from a chimney. Males that happen to fly into the plume of scent will follow it upwind. If they happen to blunder out of the plume, they presumably make a searching flight that may bring them back to it. Despite the inevitable air turbulence that distorts the shape of the plume, one of the males will probably find our female and mate with her.

Unlucky cecropia males that do not manage to locate a female before daybreak retreat to a hidden place in the foliage where they rest until dusk, when they will again make a brief flight and then settle down. About an hour before the next dawn they will yet again fly in search of females. Unmated females stop emitting the pheromone at dawn and will not move or do anything else for about twenty-three hours, until about an hour before the following dawn. Then they will again try to attract a male by releasing their sex-attractant scent. They will continue to release the pheromone for several mornings. Most of them will have coupled by the second or third night, but after four or five mornings of futile attempts, they seem to become discouraged and will, at least in captivity, lay unfertilized eggs and make little or no attempt to place them in discrete clutches or to space these clutches, as do fertilized females.

Pairs of moths that have succeeded in locating each other remain coupled until dusk, for about fifteen hours. Why they remain together for so long is a mystery; it does not take that long to transfer sperm to the female. Females that are separated from their males after much less than fifteen hours still manage to lay fertile eggs. Whatever the reason, remaining coupled at the well-hidden location of the female's cocoon does keep the nocturnal moths out of trouble during the daylight hours. If they were to separate and fly away during the day, they would probably be eaten by birds.

Shortly after dusk, the male separates from his mate and flies away. Our hypothetical female will begin to lay eggs, in clutches of about three to six, almost immediately after the male leaves. She often fastens her first clutch to the branch on which she sits, the branch to which she attached her cocoon when she was a caterpillar. After laying this first clutch, she flies for the first time in her life, seeking out shrubs and trees on which to lay the rest of her eggs. Her offspring will feed on the leaves of these plants throughout their lives as caterpillars. The moth from the dense shrub may fasten a clutch or two on the nearby birch, whose leaves nourished her when she was a caterpillar. But she will spread most of her eggs over a much wider area, some perhaps miles away from her birthplace, seldom placing more than one clutch on the same tree or shrub. After laying each clutch, she flies off to find a new site in which to lay the next one. She may occasionally return to lay on the same tree, but generally she will keep her clutches well separated from each other, thus minimizing competition for food between the caterpillars, which will eventually become huge and ravenous eaters. No one knows if the female lays eggs throughout the entire night, but we do know that she settles into a hiding place some time before daylight and continues to lay eggs on succeeding nights.

After leaving their mates, the males make another short evening flight that shuffles them among the local cecropia population. Just

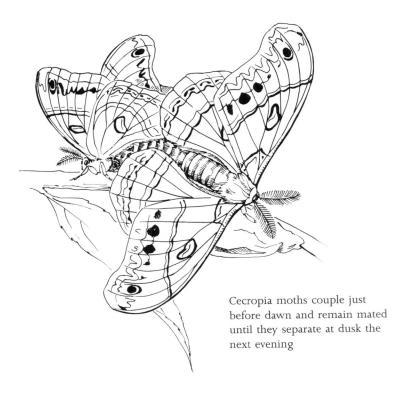

Cecropia moths couple just
before dawn and remain mated
until they separate at dusk the
next evening

before dawn of the second day, they fly in search of another
female, just as they did at the first dawn of their adult lives.
Throughout his brief life, a male continues to follow this pattern
of behavior, a short flight in the evening and a search for females
just before dawn. The female, by contrast, loses all interest in males
after she has mated only once. She will never again release the sex
attractant or couple with a male. Only virgin females are receptive
to males.

Why should there be this striking difference in the sexual
behavior of males and females in this species? The answer lies in
Charles Darwin's concept of natural selection, later to be called
the survival of the fittest. Fitness is not determined solely by an
animal's proficiency with fang and claw. As defined by evolutionary biologists, the success, or fitness, of an animal is measured by
the number of surviving progeny that it leaves behind. Some
animals, such as humans and birds, produce few progeny but

maximize their survival by investing a great deal of parental care in them. Other animals, such as some oysters, salmon, and many insects, provide little or no parental care, but instead produce very large numbers of progeny, gambling that a few of them will survive on their own. The female cecropia usually lays about 350 eggs, but, on the average, only 2 will survive to become reproducing adults. The odds against the survival of an egg of a common American oyster are far greater. The females lay as many as 60 million eggs, simply releasing them into the sea, where they may or may not be fertilized by sperm that a male of the species similarly releases into the sea. If the size of the oyster population is to stay about the same from generation to generation—the usual situation—an average of only 2 of this multitude of eggs need survive to become reproducing adults. (Just think, if only 4 out of these millions survive, the population will double in each generation, and after only ten years it will have increased to over 2,000 times its original size.) The odds against insect progeny are less astronomical, in no small part because insects or their ancestors evolved the nicety of internal fertilization. Nevertheless, most female insects lay several hundred or even several thousand eggs.

The strategy of the male cecropia is now apparent. He provides no parental care at all, but attempts to maximize his fitness by passing his genes on to as many progeny as possible by inseminating as many females as he possibly can. Cecropia is not the only species in which males adopt a strategy of promiscuousness.

The strategy of the female is different. She does provide a modicum of what could be called parental care for her eggs. Once she has been inseminated, she loses all interest in males. She is much too busy seeking out plants that her progeny will be willing to eat and then laying her eggs on them, spacing the clutches so as to avoid crowding the young caterpillars. For the rest of her short life, she spends each night distributing her eggs. Her life, as well as the life of the male, is short indeed, seldom more than five or six days, because both sexes have only a vestigial sucking tongue and cannot drink the life-sustaining nectar of flowers, as do many other moths. During her lifetime, a cecropia female may lay about 350 eggs. Since she cannot feed and lives for only a few

days, unlike many other moths, it is no wonder that she has little time to spare for males.

What we have seen so far barely begins the saga of just one insect's passage through the seasons of the year. For all animals—as well as for plants, for that matter—the achievement of evolutionary fitness involves three imperatives: the animal must avoid being eaten; it must itself eat and grow; and it must successfully reproduce itself. As with all other animals, the obstacles that may frustrate the survival, growth, and reproduction of an insect are multitudinous and varied.

Climate, weather, and the earth itself impose obstacles to survival. Insects have, however, evolved ways to survive even in the coldest climates, and some of them can survive even in the hottest and driest of deserts. Although insects are "cold-blooded" like frogs or snakes, some of them have at least partially freed themselves of the vagaries of weather. They have evolved ingenious behavioral ways to raise their body temperature and thus remain active even on cold days. The soil itself may present obstacles. Some wasps must search for the sandy soil in which they can dig the burrows in which they will raise their young. A nutrient required in an insect's diet may be in short supply in some soils and thus in short supply in the plants that the insect eats.

Other creatures are frequently obstacles to the attainment of fitness. Insects, like all other living things, exist in communities of plants and animals that interact in many ways, but principally by eating each other or competing with each other. These communities, the living constituents of the many ecosystems that encompass all life on earth, thus include organisms that, from the point of view of any one species, may be either beneficial or detrimental. Take a cecropia caterpillar, for example. From its point of view, the plants that it eats, the insects that pollinate these plants, and the birds that distribute their seeds are all beneficial. But there are also many creatures deleterious to cecropia: insects or other animals that compete with it for food and space and those that seek to destroy a cecropia by parasitizing it or by eating it.

The Most Successful Animals on Earth

We have all encountered insects, sometimes with pleasure, but sometimes with varying degrees of displeasure. In a meadow, beautiful tiger swallowtail butterflies, handsomely patterned in black and yellow, skim past us as they search for nectar-laden blossoms. Ants annoy us at a picnic. A ladybird beetle perches momentarily on the finger of a delighted child and then flies off as the child chants, "Ladybird, ladybird, fly away home . . ." A loathsome cockroach scurries across the kitchen floor in a high-rise apartment. In a garden, tiny aphids weaken a prize-winning Queen Elizabeth rose as they suck its sap. The night is cheered by the chirping of crickets and the sparkling of fireflies. A mosquito sits lightly on the back of your neck as she sucks the blood that will nourish her developing eggs. A small child, a budding naturalist exploring her backyard, is thrilled to find the plump over-wintering cocoon of a cecropia moth tucked away among the stems of a shrub.

The insects that we notice are less than the visible tip of an iceberg. For every one that we see, there are thousands of species and millions of individuals that we do not see. Some are so small that they are seldom noticed. Indeed, their small size is one of the reasons that insects have been so successful. Other insects are active only in the dark, and many are hidden in the soil, in the tissues of plants, or in the bodies of the insects or other animals that they parasitize.

Few of us realize that the insects, as a group, constitute the greatest evolutionary success story of all time. They have persisted for hundreds of millions of years, occur in almost endless variety, and are found—usually abundantly—virtually all over the world. They are indispensable citizens of almost all of the ecosystems that constitute all life on the continents: lakes, rivers, meadows, forests, pastures, crop fields, orchards, and others. Among these are all of the ecosystems, other than those in the oceans and seas, from which we humans—even those of us who ride subways and walk on concrete—obtain the food without which we could not live.

The ultimate indicator of the evolutionary success of insects is their importance to life on earth, which results from their interactions with other living things within ecosystems. But before examining the difficult question of the important roles that insects play in ecosystems, we should consider the more obvious and, relatively speaking, more easily measured signs of their success. One can express with numbers or maps the millions of years that insects have endured on the geological time scale, how many different species have evolved, their worldwide distribution, and the magnitude of their populations. An understanding of the complex and subtle ways in which insects interact with other animals and plants is far more elusive, and we will come to that later in this chapter.

About 40 million years ago, insects were trapped in sticky resins that oozed from trees in northern Europe. Today, fossilized globs of these resins wash up on the beaches of the Baltic Sea. This resin is the amber used in jewelry. A few pieces of amber contain beautifully preserved insects, but they are almost identical to insects that we see around us today. To see truly primitive and ancient insects, we must go much further back in time. In the fossil record, the first wingless insects appeared 400 million years ago, about 20 million years after their arthropod ancestors left the oceans to invade the land, and long before the first reptiles, birds, or mammals had evolved. About 315 million years ago, 90 million years before dinosaurs walked the earth, primitive winged insects flew in the great forests of giant horsetails, club mosses, and tree ferns. By 65 million years ago, these primitive

plants had declined and flowering plants dominated the continents, providing new resources for the hordes of more modern
insects that co-evolved with the plants and now eat their nectar,
pollen, leaves, or other tissues. Like the insects in the Baltic amber,
most of the insect fossils from this relatively recent period are
similar, if not virtually identical, to insects that live today. In
response to this great new abundance of insects, there evolved
many insect-eating vertebrates, especially the passerine birds—the
songbirds—which include over half of all the bird species, among
them such highly insectivorous groups as the flycatchers, swallows, vireos, and warblers.

Today most of the animal species on earth are insects. Over 1.2
million different kinds of living animals are known to science.
About 900,000 of them, 75 percent of the total, are insects. The
beetles alone number about 250,000 species, almost as many as
all of the animal species other than insects put together. The
beetles outnumber all of the vertebrates in the aggregate, the
familiar fish, amphibians, reptiles, birds, and mammals, by a
factor of about five to one. When the great British population
geneticist J. B. S. Haldane, was asked what could be learned about
the Creator from studying His creations, Haldane replied that the
Creator seems to have an "inordinate fondness for beetles." How
did Noah ever manage to supervise the boarding of so many
beetles onto the ark?

Millions of animals, most of them insects, are still unknown to
science. Conservative estimates indicate that there may be about
9 million species of insects that have yet to be discovered. A much
larger but possibly reasonable estimate is that about 30 million
species of insects remain unknown. Terry L. Erwin of the Smith-
sonian Institution in Washington based this estimate on the number of previously unknown insect species that he found in South
America by making exhaustive collections from a previously almost unexplored habitat, the leafy forest canopy from one to two
hundred feet above the ground.

The vast majority of undiscovered insects live in the tropical
forests, particularly those of South and Central America. But these
forests are being destroyed rapidly, mostly by being converted to

farms or pasture land, much of which will be exhausted and useless after a few years. If the destruction continues at the present rate, these forests will be gone by the first decades of the twenty-first century, leaving only scattered, small remnants that can sustain little species diversity. Animals and plants are threatened with extinction everywhere in the world, not only in the tropics. In the United States alone, thousands of plants and animals are threatened with extinction. Considering only the insects, over 700 species are listed as being endangered or threatened or are candidates for those lists. What wonderful things might we learn from these unknown insects, other animals, and plants that are about to go extinct? Of what practical use might they be? Terry Erwin asked, "Do we not need to know, in these days of genetic engineering, what the world holds in the way of genetic diversity?"

In the tropics and the temperate zones, insects are abundant virtually everywhere on land and in fresh water. But no insects live on the polar ice caps, they are extremely rare in the oceans and seas, and only a handful of species occurs in the small areas of Antarctica that are not perpetually covered by ice. In the Arctic they are surprisingly abundant. Bumble bees, butterflies, and beetles live as far beyond the Arctic Circle as plants grow. On the Arctic tundra, dense swarms of mosquitoes plague caribou, birds, people, and virtually any vertebrate that they can find. Some insect species have even adapted to habitats that seem inhospitable to life. Some live among rocks and snow 20,000 feet above sea level in the Himalayas. Maggots of the aptly named petroleum fly feed and swim about in pools of crude oil. A few insects live in hot springs where the temperature climbs to 120° F. Deep in caves where light never reaches, there are beetles and cave crickets, many species eyeless and ghostly white in color. In the Great Salt Lake of Utah, a lake so salty that no fish can survive in it, brine flies are so abundant that the Indians once gathered basketfuls of their tiny pupae for food.

Mass outbreaks come to mind when we think about the magnitude of insect populations. In the soil of an infested corn field there may be 25 million click beetle grubs per acre. A single swarm of migratory locusts (grasshoppers) in North Africa may consist of 8 billion individuals and weigh 20,000 tons. With Moses as his spokesman, Jehovah visited ten plagues upon the ancient Egyptians who held Israel in bondage. The third and fourth were plagues of lice and flies, but, according to Exodus 10:15, the eighth plague was a swarm of locusts that "covered the face of the whole earth, so that the land was darkened; and they did eat every herb of the land, and all the fruit of the trees . . . and there remained not any green thing, either tree or herb of the field, through all the land of Egypt." In the 1980s similar swarms of locusts devastated parts of North Africa and western Asia, and in 1993 vast locust swarms wreaked havoc in southern Pakistan. Even North America once had its own species of migratory grasshopper. As reported by Curtis Sabrosky, in the 1870's, great swarms of Rocky Mountain grasshoppers, which fortunately seem to be extinct or inactive now, migrated across the Great Plains. An eyewitness in Nebraska, a surveyor, reported that one migrating swarm was 100 miles wide, 300 miles long, half a mile high, and included at least 124 billion grasshoppers. This huge swarm left fields as barren as if they had been burned: only holes in the ground showed where herbaceous plants had stood; trees were stripped of their leaves and green bark; and the dry straw was eaten from brooms.

The settlers on the Great Plains did not eat the locusts that plagued them, although that would have been the ultimate revenge. Although Europeans and non-native Americans have persistent prejudices against eating insects, the members of most non-Western cultures eat insects of many kinds. For example, the people of Africa and the Mideast—not about to pass up an abundant and protein-rich source of food—harvest the locusts that so often overrun their lands. Deuteronomy 14:19 warned the Israelites that "all winged creeping things are unclean to you: they shall not be eaten." But Leviticus 11:20 specifically exempted locusts: "Yet these may you eat of all winged creeping things . . . [those] which have legs above their feet, wherewith to leap upon the earth . . . the locust after its kind, and the bald locust after its kind, and the cricket after its kind, and the grasshopper after its kind."

Population outbreaks are dramatically obvious, but normal populations, which often go unnoticed, may also be extremely large. There may be almost 30,000 insects, mostly eaters of dead organic matter, per square yard in the ground litter and the top one inch of the soil in a deciduous forest in eastern North America. One termite nest in Australia was found to contain 1,806,000 individuals. A queen honey bee lays about 200,000 eggs each year, and at any given moment a hive may contain as many as 55,000 worker bees. In a weedy, abandoned field in North Carolina, the plant-feeding insects alone—leaving out all parasitic, predaceous, and scavenging species—weighed in the aggregate nine times as much as all of the much larger and far more conspicuous sparrows and mice combined. Samples taken on an East African plain showed that two species of ants—only those two among the many hundreds of insect species present—numbered about 10 million per acre and were roughly equal in weight, on a per-acre basis, to the combined weight of the large grazing animals of the plain—zebras, antelopes, wildebeests, and others.

The world population of insects—all individuals of all species combined—is astronomical and few scientists have had the temerity to even guess at its magnitude. In his delightful 1968 book, *Six-Legged Science*, the late Canadian entomologist Brian Hocking presented an educated guess, that about 1 quintillion insects (1 followed by 18 zeroes) inhabit earth. Using a conservative estimate of the average weight of an insect, he calculated that the insects of the world (2.7 billion tons of them) weighed twelve times as much as the then current world population of 3 billion people. In the twenty-five years since the publication of Hocking's book, the human population has swollen to 5.7 billion, but as far as we know the insect population has remained about the same. If Hocking's estimate is correct, the insects now outweigh us by a factor of less than seven. Between seventy and one hundred years from now, the human population will have grown so large that it will nearly equal the insects in weight, if we make the unlikely assumption that there will be no decline in insect numbers. Can the ecosystems that will have to produce the food for those 2.7 billion tons of humanity function with such a low insect-to-human ratio?

The plants and animals in an ecosystem form an integrated and interdependent community of organisms (not unlike Adam Smith's vision of a society in which farming, manufacturing, merchandising, and all of the other occupations form a functioning system coordinated and guided by free competition in the marketplace). Green plants capture the energy of the sun and make it available to other living things in the form of sugars. Insects and, to a much lesser extent, other animals such as hummingbirds and bats are the pollinators that make possible the existence of most plants. Plant populations are prevented from expanding to ruinous numbers by competition between and within them, by disease, and by the insects and other animals that eat them. Insect populations do not increase catastrophically because of competition between and within them and because of disease and the parasites and predators, many of them other insects, that eat them in turn. Scavengers such as bacteria, insects, and vultures break down and return to the soil dead organic matter such as leaves, feces, and animal carcasses, whence their atoms and molecules will be resurrected to form the bodies of succeeding generations of plants and animals.

Recognizing that insects are important or even indispensable to an ecosystem is tantamount to recognizing their evolutionary success. Let us consider one example. Just as the communications industry is essential to today's society—imagine life without a postal system, telephones, e-mail, or fax machines—pollinating insects such as bees, flies, and moths are essential to terrestrial ecosystems. About 80 percent of the flowering plants are partially or wholly dependent upon insects to carry their pollen from the male parts of a blossom to the female parts of another or the same blossom. Without pollinating insects, tens of thousands of different kinds of flowering plants would become extinct or exist only as greatly reduced populations. The continents would be domi-

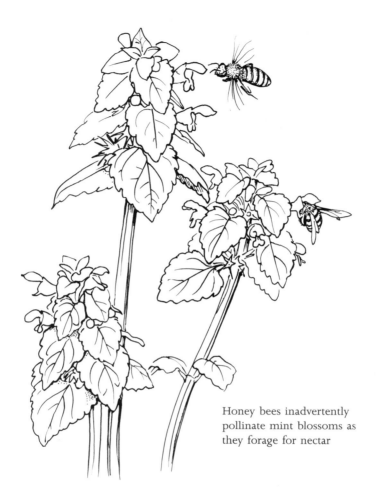

Honey bees inadvertently
pollinate mint blossoms as
they forage for nectar

nated by grasses, pines, firs, and a few other wind-pollinated plants such as cottonwoods and ragweeds. There would be no chocolate or coffee since cacao trees and coffee trees are pollinated by insects. Considering only sweet fruits that people eat, without pollinating insects there would be no melons, figs, peaches, plums, apricots, cherries, strawberries, raspberries, blackberries, blueberries, cranberries, kiwis, citrus fruits, pears, or apples.

Now let us look at some events that demonstrate specific examples of ecological interdependence and the importance of the roles of insects. Charles Darwin is usually credited with the whimsical notion that old maids were the mainstay of the once great British Empire. The argument goes that old maids kept cats, that cats caught the field mice that destroyed bumble bee nests, that bumble bees pollinated clover, that clover made good beef, and that good beef made the brave soldiers who extended and defended the Empire. In a more serious vein, Darwin argued that red clover would go extinct in England if there were no bumble bees to pollinate it. (Darwin overstated his argument. Bumble bees are indeed the major pollinators of red clover because their long tongues are ideally suited to sipping nectar from its deep blossoms. But other insects do sometimes pollinate red clover.) To prove his point, Darwin covered a patch of blossoms with a net to exclude bumble bees and found that the covered plants did not produce a single seed. Since the net excluded other possible pollinators too, Darwin's experiment did not prove that bumble bees are the only pollinators of red clover in England. But it did prove that this plant cannot reproduce itself unless it is pollinated by insects of some sort. Late in the nineteenth century, only a few years after Darwin's death, an event in New Zealand demonstrated conclusively that red clover depends upon insects for pollination, and that it produces abundant seed even if there are only bumble bees to pollinate it. Red clover is not native to New Zealand, but farmers there imported seed from abroad and grew the clover to feed their livestock. Importing seed was expensive but necessary. Red clover grown in New Zealand produced very little seed, apparently because none of the native insects are capable of pollinating it. Honey bees, first introduced to New Zealand in 1839, were probably responsible for the minimal pollination that did occur. In 1885 bumble bees were successfully introduced into New Zealand. They soon became established, and to this day they pollinate red clover so efficiently that it bears abundant seed in New Zealand.

An ecosystem is normally in a rough state of balance. Ups and downs occur, but seldom does a plant or animal population increase uncontrolledly to disrupt the system. Virtually all plants

and animals produce more offspring than necessary, like the oysters that I discussed earlier. If only 2, or 0.04 percent, of the 500 or so eggs produced by a pair of house flies survive to become reproducing adults, the parents will have replaced themselves, and the population will remain stable from generation to generation. If 5, only 1 percent, of them were to survive, the house fly population would increase by an ecologically, societally, and economically calamitous 250 percent about once every three weeks. Fortunately, competition, disease, unfavorable weather, and other factors do help to hold populations of house flies, oysters, and other organisms in check. But without the regulating effect of parasitic, predaceous, and plant-feeding insects, many animal and plant populations would expand out of control. This is difficult to prove, but the ways in which entomologists alleviated two ecological disasters brought on by human activity are essentially unintentional experiments that prove, on a grand scale, that insects play essential roles in maintaining population balances.

One of these disasters began in 1868—just after the Civil War and eight years before Sitting Bull defeated Custer at the Little Bighorn. A tiny Australian insect that sucks sap from trees suddenly appeared in southern California. No one knows how this insect crossed the Pacific Ocean, but it was probably brought in accidentally on plants imported from Australia. By 1886 this insect, the cottony cushion scale, had spread widely, had infested many kinds of plants, and had killed thousands of orange trees. The scale seemed to have no natural enemies in California. The state's infant citrus industry was doomed unless a remedy could be found.

Charles V. Riley, first director of the Division of Entomology of the U.S. Department of Agriculture, realized that it might be possible to control this plague of immigrants from Australia by importing some of their natural enemies from that distant continent. In its native Australia, the cottony cushion scale is never abundant enough to be a major pest. But Riley was barred from using USDA funds to send an entomologist to Australia to search for parasites and predators of the cottony cushion scale. (Congress had put a rider prohibiting foreign trips by employees on the appropriation bill for the USDA. The rider was specifically aimed at Riley himself, who made frequent junkets to Europe at government expense.) But a way around this difficulty was found. In 1888, Albert Koebele, a USDA entomologist, was attached to the American delegation to an international exposition in Melbourne—at a total cost of about $1,500—ostensibly to represent the State Department but actually to search for enemies of the cottony cushion scale.

On October 15, 1888, Koebele found a small ladybird beetle, the vedalia, eating cottony cushion scales in a garden in Adelaide. During the next three months he sent 129 of these beetles to California. They were placed under a tent covering an orange tree infested with the scales. The vedalias multiplied and killed nearly all of the scale insects on that tree. When the tent was opened, the beetles spread to nearby trees and freed them of infestation. Vedalia beetles from these trees were then given to orange growers all over southern California. By 1890 the plague was over, the little vedalia ladybird was acknowledged to be an "entomological miracle," and Koebele and Riley were hailed as heroes. To this day, populations of cottony cushion scales, not large enough to be destructive, are held at this low level by the correspondingly low populations of vedalias with which they coexist in California.

The importation of prickly pear cactus into Australia late in the eighteenth century eventually resulted in another ecological calamity of far more awesome proportions. Grown as garden plants, the cacti flourished and soon escaped from cultivation. By 1900 they had overrun almost sixteen thousand square miles of range land, an area about twice the size of New Jersey. In 1925 they covered about ninety-four thousand square miles, an area almost twelve times the size of New Jersey, and it was obvious that they would continue to spread. All of this land was practically useless, and about half of it was so densely overgrown with tangles of the spiny cacti that it was literally impenetrable to kangaroos,

Vedalia beetles attack
cottony cushion scales

people, and cattle. In the Western Hemisphere, the only part of
the world where cacti occur naturally, nothing even approaching
this plague of cacti had ever been seen. Held in check by their
natural enemies, which do not occur in Australia, New World
prickly pear cacti grow in scattered clumps, never in huge, impenetrable tangles.

As recounted by Paul DeBach, the plague of cacti in Australia
was eliminated by bringing in several cactus-feeding insects from
the New World. The most effective was a moth from South America without a common name but with an appropriately descriptive
generic name, *Cactoblastis*. These insects were introduced in 1925,
and by 1937 they—but especially the *Cactoblastis* caterpillars—
had destroyed the last dense stand of cactus. To this day, the cacti
in Australia—their populations still controlled mainly by *Cac-*

toblastis—grow only in scattered clumps, and cattle, sheep, and kangaroos graze on once useless land.

Just as a city without sewage and trash disposal would soon choke on its own waste, an ecosystem cannot persist without scavengers, the decomposers of dead organic matter. Although bacteria and fungi are the ultimate decomposers, insects are also important. Just how important they can be is shown by another unintentional experiment on a grand scale—once again in Australia—interestingly described by D. F. Waterhouse, an eminent Australian entomologist, in the April 1974 issue of *Scientific American*. Certain scarab beetles use the large, moist dung pats of buffalo and other cowlike grazers as food for themselves and their young. The sacred scarab of ancient Egypt, for example, provides for each of its young by forming a large ball of cow or buffalo dung that it rolls to a suitable site and buries in the soil along with one of its eggs. The scarabs of Australia, however, are evolutionarily adapted to cope with only the small, dry pellets of dung produced by kangaroos, the only large grazing animals that occur naturally on that continent. When the English brought cows to Australia, there were no scarabs capable of decomposing their dung. Cow droppings accumulated, remaining on the ground for months or even years before decomposing, seriously impairing the productivity of pastures and range land by choking off the plants on which cows feed. According to Waterhouse, the 30 million cows of Australia deposit about 360 million pats of dung per day. As you have probably guessed, this problem was solved by introducing into Australia from Africa and Europe several species of scarabs that were evolutionarily adapted to cope with the large, moist dung pats of cows.

These unintentional ecological experiments show that ecosystems must include other as yet unsung insect heroes—unsung because we are not aware of their essential roles in these ecosystems, unsung because we have never seen what would happen if they were not there. If the vedalia beetle were exterminated from its native Australia, would the cottony cushion scale remain a small and relatively benign presence there? Could the cattle that now live on the Great Plains of North America survive if there were no insects to help decompose their dung, insects that evolved to exploit the dung of the immense herds of bison that once wandered those vast grasslands? Would red clover exist anywhere on earth if there were no bumble bees to pollinate it? What would happen if we were unlucky enough to synthesize a new insecticide with which we unintentionally exterminated the cactus-feeding insects in Texas? Cacti would overrun the state, crowding out many grasses and other plants. Deer, cattle, and pronghorns that eat these other plants would probably not survive. Insects that eat these plants would also disappear, as would the other insects, birds, and small mammals that eat the plant-feeding insects.

There is no doubt that insects are critically important to many of the world's ecosystems. On that score, they are certainly an extraordinary evolutionary success.

Insects are not just biological entities to be appreciated from only a scientific point of view. Humans—at least some humans—perceive them esthetically and philosophically. Insects—some perhaps more so than others—exemplify the excellence and the beauty that spring from the evolutionary process. The eternally ongoing process by which the Prime Mover, Nature, God if you will, creates species. In Joseph Conrad's Lord Jim, Stein, the German entomologist, ponders a rare and beautiful butterfly: "Marvelous! Look! The beauty—but that is nothing—look at the accuracy, the harmony. And so fragile! And so strong! And so exact! This is Nature—the balance of colossal forces. Every star is so—and every blade of grass stands so—and the mighty Kosmos in perfect equilibrium produces—this. This wonder; this masterpiece of Nature—the great artist."

Finding and Courting a Mate

Most animals must at some time in their lives locate and identify appropriate mates. As we already know, American oysters are an exception. Males and females simply release their sperm and eggs into the sea, thus avoiding the need to locate mates and recognize them as members of their own species and as members of the opposite sex. We humans are seldom conscious of these problems. When we reach reproductive maturity we have no difficulty in finding members of the opposite sex or in recognizing our endearing opposites as humans. We have, after all, been surrounded by other people all of our lives. But consider the cecropia moth that has just emerged from its cocoon. Its mother abandoned it as an egg and it has never seen an adult cecropia. How is it to recognize another cecropia as a member of its own species, let alone as a member of the opposite sex? How will it even find another cecropia? The nearest company may be a mile or more away.

As we have already seen, these problems are solved by the sex-attractant pheromone secreted by the females. This pheromone conveys three essential messages to the male who smells it: I am also a cecropia; I am a female cecropia; find the source of this odor and you will find me. The ability to understand these messages, a capability that the males display even at the first dawn of their adult lives, is obviously inborn—encoded in the genes. After all, the males have not had an opportunity to learn the meaning of the pheromone. Sex-attractant pheromones are usually emitted by females, but in quite a few insects, including the infamous boll weevil, it is the male that attracts the female by his scent.

Before I go on to discuss how insects find and court mates, I must digress to consider metamorphosis and how it relates to mate-finding and reproduction.

Cecropia, all of the other moths and butterflies, the aphidlions and their relatives, the beetles, the flies, the fleas, and the wasps and bees undergo what entomologists call complete metamorphosis. These insects pass through four different life stages: the egg; the larva or growing stage, more specifically referred to as the caterpillar, grub, or maggot; the pupa or transformation stage, called the chrysalis by butterfly collectors; and the adult or reproductive stage. The transformation that occurs in the pupal stage has made it possible for larvae and adults—caterpillars and moths, maggots and flies, grubs and beetles—to follow different evolutionary paths that have produced two different life stages with radically different anatomies and behaviors.

Imagine that an extraterrestrial biologist visits earth from a distant planet where metamorphosis is unknown. Now suppose that on a day in late June he or she (or possibly it, since sex and gender might be unknown on that faraway planet) happens to collect a cecropia moth and a cecropia caterpillar. This alien being would surely think that the moth and the caterpillar are unrelated animals as different as birds and snakes. Small wonder, the caterpillar and the moth *are* very different. The caterpillar has a naked, wormlike body; no wings at all; six short legs on the thorax and, on the abdomen, ten stubby prolegs good only for clutching the stems or leaves of plants. Its stout jaws chew up the huge quantity of leaves that it must eat to support its growth and to make the fat that it will store for later use. Cecropia moths, by contrast, have compact, hairy bodies; enormous wings covered with tiny scales; and only six long legs that serve as landing gear. Although cecropia moths have vestigial mouthparts and do not feed, most adult insects do feed. Other species of moths and butterflies have a long, thin tongue, or proboscis, like a flexible drinking straw. It can be coiled out of the way beneath the head or extended to suck nectar from blossoms.

The different life styles of caterpillars and moths or other larval forms and adults are tailored to serve the special functions of these two life stages. The raison d'être of the caterpillar is to eat, grow, and survive. It is not much more than a digestive system on a caterpillar tread. The raison d'être of the moth, which does not grow, is to survive and reproduce. Moths and butterflies have been referred to as winged gonads or, as Carroll M. Williams said, "flying machines devoted to sex." They fly long distances to seek mates or to lay eggs, thus requiring abundant energy. The nectar that most of them drink lacks many of the nutrients required for growth, but since it consists mostly of sugar and water, it is a rich source of energy to fuel the muscles that drive the wings.

Now let us imagine that the extraterrestrial biologist collects an adult grasshopper and a partly grown grasshopper whose wings are not yet fully developed. Although this biologist is a stranger to the earth, he, she, or it would certainly realize that the immature form is a baby grasshopper. Recognizing that immature and adult grasshoppers are the same kind of animal requires little perspicacity since the two are very similar.

Grasshoppers and other insects such as dragonflies, cockroaches, crickets, bed bugs, aphids, and cicadas experience *gradual metamorphosis*. There are only three life stages: the egg; the growing stage, called the nymph to differentiate it from a larva; and the adult or reproductive stage. Since there is no pupa, or transformation stage, nymphs and adults behave and look very much alike except for their size and the state of development of their wings. Nymphal and adult grasshoppers, for example, live in the same habitat, and both use their jaws to chew the leaves of plants and tend to be active jumpers. Except for the fully developed wings and genitalia of the adult, neither life stage has become further specialized for either growth or reproduction. The body form and life style shared by nymphs and adults have evolved as a compromise that must serve both functions.

Considering the benefits of specialization, it is not surprising that complete metamorphosis is an evolutionarily more successful

A newly molted dog day cicada sits above the nymphal skin that it shed just a short while ago

strategy for survival than is gradual metamorphosis—at least as judged by the numbers of species in each category that have survived to the present day. Of the known insects species, only about 135,000, or 15 percent, have gradual metamorphosis. There are, however, over 765,000 insect species that have complete metamorphosis, about 85 percent of all the known kinds.

To return to the problem of finding a mate, students sometimes ask me how powerful the cecropia sex-attractant pheromone is. A good question, but one that needs to be re-phrased lest we put the cart before the horse. We should ask instead about the sensitivity of the male to the pheromone of the female. Not much detailed information is available for cecropia, but it is available for the common silk moth, whose caterpillar stage produces the silk in our blouses and neckties. Cecropia males probably respond in about the same way. According to reasonable estimates, the male silk moth can perceive and respond to a few hundred molecules of pheromone in a cubic centimeter of air. (A cubic centimeter is about four-tenths of an inch on a side.) That small quantity of air contains about 25 quintillion molecules—the number 25 followed by a string of 18 zeros. The male moth can thus perceive a few hundred molecules of pheromone dispersed among almost 5 quintillion molecules of oxygen, almost 20 quintillion molecules of nitrogen, and small amounts of carbon dioxide and other gases. The male silk moth is, indeed, very sensitive to the pheromone produced by the female.

The scent receptors of moths are on their antennae. The antennae of females, which do not need to respond to ultra-low concentrations of a sex-attractant pheromone, are usually much smaller than those of males. Cecropia moths are no exception. The size of the antennae distinguishes the sexes at a glance. The featherlike antennae of the males are eye-catchingly ample, much more broadly widened than are the relatively modest antennae of the females. Thousands of tiny scent receptors, each one of which can probably detect a single molecule of the sex-attractant phero-mone, are accommodated by the enlarged antennae of the male. The excitation of only a few hundred of these individual receptors is probably sufficient to alert the male to the presence of the pheromone, and to initiate the complex pattern of behavior that begins with upwind orientation and culminates in mating.

From the point of view of cecropia's natural history, it is particularly interesting to know the maximum distance from which males can locate females. This distance is one of several factors that set a limit on how small a cecropia population can get and yet survive—how widely dispersed the moths can be and yet locate each other for the purpose of mating. This maximum

distance may be much longer than the usually shorter distance over which a male can actually smell the pheromone. As the scent drifts down wind, the pheromone becomes more and more diffuse and the molecules of pheromone eventually become too dispersed for even the highly sensitive males to perceive—much as a plume of smoke becomes more and more dispersed as it drifts downwind. If we could follow a male on the journey that takes him to a female, we would see, as postulated by Edward O. Wilson and W. H. Bossert, that it generally consists of two parts: first, random movement, not directed by the perception of the pheromone, that may by luck bring him closer to a female; then more or less directed upwind flight along the perceptible portion of her pheromone plume, which will ultimately bring him into contact with her.

At least for cecropia and the other giant silk moths, this distance tends to be very great, much greater than even our most optimistic notions of the distance from which a male can smell a pheromone. Early in this century, Phil and Nellie Rau, intrepid St. Louis naturalists, recaptured, in a trap baited with a female, over 11 percent of the cecropia males that they had released about three miles away. Many years later, my colleagues and I recaptured one male almost eight miles away from its release point and several other males over four miles from their release point. More males would probably have been recaptured from these distances if they had not flown through territory replete with wild, pheromone-releasing females that were competing with the bait females in our traps.

Promethea males, giant silk moths closely related to cecropias, are known to fly even longer distances, but in this case we released the males farther away and the females that baited the traps had little or no competition from wild females. One male promethea captured in a trap had flown at least nine miles in three days. Another one, which may hold the distance record among moths, located the bait female by flying almost twenty-three miles in three days. All of these are straight-line distances. The moths' flights were probably wind-assisted, except for the relatively short distance over which they flew upwind within a pheromone plume, but the actual routes that they took may have been circuitous and much longer than the straight-line measure.

Tens of thousands of insect species use sex pheromones to attract mates from a distance, including many different kinds of cockroaches, termites, scale insects, beetles, flies, bees, wasps, and moths. Celibate male cockroaches are driven to a sexual frenzy if a piece of paper on which virgin females have walked is put into their cage. The fossil lineage of cockroaches goes back to the Upper Carboniferous, the great age of coal formation over 300 million years ago. These insects may well have used sex-attractant pheromones that far back in time just as they do today.

Some parasites and predators of insects have broken the pheromone code of other insect species. They have evolved to recognize and follow the sex-attractant pheromones of the hosts that they parasitize or of the prey that they eat. Some tiny wasps are strongly attracted from a distance to certain pheromone-releasing scale insects, and they lay their eggs in the tiny bodies of these scale insects, where their larval offspring feed and grow as parasites. A predaceous beetle that eats bark beetles orients from far away to trees infested by its prey and can be caught in traps baited with nothing but a synthetic form of the sex-attractant pheromone produced by these bark beetles.

Synthetic sex-attractant pheromones have even become weapons in humanity's battle against pest insects. They are used to bait traps that monitor the rise and fall of pest populations, thus indicating whether or not a population is large enough to pose a threat. They can even be used to control pest populations. When the synthetic attractant of the pink bollworm, one of the most destructive pests of cotton, is released into the air in cotton fields, the males fail to find and fertilize the females, presumably because the pheromone signals released by females are lost among a multitude of false signals. This use of the harmless pheromone is as successful in reducing an infestation as is the application of an insecticide that may be ecologically harmful or even a direct threat to people.

There are many pheromones other than those that attract mates from a distance. Some facilitate behaviors or physiological func-

tions involved with sex and reproduction in other ways. Male butterflies, close relatives of moths, first locate potential mates by sight, but their vision is not acute enough to distinguish females of their own species from other butterflies. Final identification is made at close range via a pheromone produced by the female. The male, in turn, releases a pheromone that identifies him to the female and that may act as an aphrodisiac. When a male cecropia finally arrives at the source of the pheromone plume, he lands close to the female and almost immediately makes genital contact. To a human it looks like rape. But her release of the attractant pheromone may constitute implied consent, and upon his arrival he possibly releases a pheromone that informs her of his identity and intention, although there is no experimental evidence to prove this point for cecropias. Some male cockroaches, and some crickets too, exude on their bodies a substance on which the females nibble during copulation. This substance, waggishly named seducin, is an aphrodisiac pheromone. The social integrity and hence the survival of honey bee colonies, which consist of tens of thousands of individuals and persist for years, depends upon the queen's maintaining a monopoly on reproduction. She protects her monopoly by continually producing pheromones that inhibit the development of the ovaries of worker bees, which are all females and constitute almost the entire population of the colony.

Insects are by no means the only animals that communicate with each other through pheromones. Although first discovered and most thoroughly studied in insects, these chemical signals, we now know, are also used by many different kinds of amphibians, fish, reptiles, and mammals. Birds are notably absent from the list of pheromone producers. Never noted for the keeness of their sense of smell—except for a few species, including kiwis, a vulture, and some sea birds such as petrels, albatrosses, and shearwaters—they seem to communicate mainly by voice and visual displays. Many mammals use pheromones, often carried in the urine, to mark the boundaries of their territories. Witness the behavior of dogs at telephone poles and fire hydrants. Male house mice make a pheromone, carried in the urine, that markedly speeds up the estrus cycle of females. The females are thus in breeding condition and sexually receptive more often than they would be in the absence of a male. The male himself need not be present. A tiny drop of his urine applied to the female's bedding each day by an experimenter is enough to accelerate her estrus cycle.

Women, experiments have shown, are similarly affected by a pheromone that can be extracted with alcohol from the hair tufts in the armpits of men. (The men showered daily but used neither deodorant nor cologne.) Three times a week a small amount of this extract was rubbed on the upper lips of several women with a history of very irregular menstrual (estrus) cycles. They were asked not to wash their faces for at least six hours. These women reported no odor from the extract, but their menstrual cycles became significantly more regular. A control group of similarly irregular women was not affected by the application of plain alcohol.

Quite a few people think that humans produce other pheromones that are aphrodisiacs which trigger and maintain sexual behavior. There is abundant anecdotal evidence, but the scientific study of human pheromones is in its infancy and there is only a smattering of experimental evidence. A prominent European animal behaviorist once told me that he had great success seducing women if he danced with them closely while he wore in his breast pocket a handkerchief that he had rubbed in his recently washed but deodorant-free armpit. It is worthy of note that, of all the mammals, only humans, gorillas, and chimpanzees have odor-secreting glands in their axillary areas (armpits).

Humans are not known to secrete musks, which serve as the sex pheromones of many other mammals, although humans are known to secrete some of the chemical components of musks. In Western culture both women and men expropriate the pheromones of other species by using perfumes and colognes that incorporate musk from deer, beaver, or civet cats. (My wife used to tell our teenage daughters that musky perfumes are too "sophisticated"—read too sexy—for young girls.) The axillary areas, vagina, and penis have all been implicated as sources of human

sex pheromones. In a controlled test, male and female college students wore T-shirts for twenty-four hours without bathing or using deodorant. Each T-shirt was then hidden in a container. When asked to sniff holes in these containers, the great majority of the students were able to recognize their own odor and also to identify the sex of the person who had worn each T-shirt. None of the twenty-nine students involved in the test reported the odor of the T-shirts to be objectionable. In another experiment, a chair in a dentist's waiting room was sprayed with a chemical compound that occurs in the armpit secretions of men. Women tended to prefer that chair but men tended to avoid it.

Pheromones play significant roles in behaviors other than courtship and sex. Female fruit flies protect their maggot offspring from crowding by marking each fruit in which they lay an egg with a pheromone that warns other females not to lay an egg in the same fruit. Ants use a pheromone to mark a scent trail that other ants can follow from the colony to a source of food. Some aphids release an alarm pheromone when they are attacked by a predator, thus warning other aphids of the predator's presence. When a worker honey bee stings an intruder at the colony, perhaps a bear or a human, she also releases an alarm scent that incites other workers to join in repelling the intruder.

Some insects use sounds in courtship. Louis M. Roth was the first to demonstrate conclusively that male mosquitoes locate females by the sound that they make in flight, a sound that is also audible to the human ear. Most of us have been kept awake in the dark by the droning of a female mosquito as she flies about in search of a victim from whom she can take a blood meal. It is that sound that attracts males. Roth proved his point by striking a tuning fork behind a cloth in a cage full of virgin male mosquitoes. If the pitch was right, like that of a female, the males clustered on the cloth as close as they could get to the vibrating tuning fork. They grasped the cloth much as they would grasp a female; some of them even tried to copulate with the cloth or with each other.

The antennae of male mosquitoes, far hairier and bushier than those of the females, act as ears, their sensitivity to sound waves enhanced by the long, thin hairs that extend out from the shaft of the antenna. Under normal conditions, a male attracted by the sound of a flying female does not copulate with her until after he has stroked her body with his legs. Taste receptors on his legs perceive a pheromone that identifies her as a member of his own species.

Among the other insects that use sound in locating and courting mates are the grasshoppers, the katydids, and the crickets. The songs of grasshoppers are heard mainly in daylight but katydids and crickets are heard mainly at night. Some katydids sing their name: "Katy did; Katy didn't." To us the chirps of crickets are just cheerful sounds in the night, but for the crickets they are inextricably involved with the race to find a mate and to be survived by their own offspring.

Crickets, like grasshoppers and katydids, chirp by stridulating, an action similar to rubbing a file against the edge of a piece of sheet metal or running the thumbnail along the teeth of a comb. The females of most species of crickets are mute because they lack stridulatory organs, but they, like the males, have a pair of ears, one located on each foreleg. Male crickets chirp by rubbing their front wings together. Each wing has a filelike structure that rubs against a scraper on the opposite wing. The chirps of some crickets are amplified because they sing from the mouths of tunnels that they dig in the shape of megaphones or bandshells. Other crickets sing in the open air with no intensification of their songs.

The rate at which crickets chirp varies with the air temperature—the warmer the night, the faster the rate of chirping. With some crickets, among them the snowy tree cricket, sometimes called the thermometer cricket, the relationship between temperature and calling rate is so predictable that one can calculate the temperature from their calls. Counting the number of chirps heard in thirteen seconds and adding forty to that number is a reasonably accurate estimate of the air temperature in degrees Fahrenheit.

Some crickets sing three different types of song. First, there is a loud aggressive song. It proclaims that the singer is in possession of his territory, and that he will defend it against intruding males. If there is an intrusion, the two males first try to settle their conflict by outsinging each other. This often settles the conflict, and usually in favor of the original holder of the territory. Sometimes the two males go beyond a simple chirping match. They may resort to lashing each other with their antennae, to sparring with their forelegs, or even to dismembering each other.

The Chinese have been betting on cricket fights for centuries. The Chinese term for a cricket fight is tou chiür chiür, the last two syllables an onomatopoeic rendering of the cricket's chirp. During the cultural revolution in Red China, cricket fighting was discouraged, but in more recent years it has again become extremely popular. Two male crickets, each selected and valued for his fighting ability, are brought together in a small cage made of straw or splinters of bamboo. Since there is no hiding place in the cage, the males clash, usually beginning with a chirping match. But if they do not soon move on to physical combat, their handlers incite them to violence by stroking the tips of their antennae with a fine straw.

A second and different loud song calls in females. (In 1913 a German researcher showed that the male's calling song alone will attract females without the intervention of visual or chemical stimuli. He induced males to sing to females over a telephone.) When a female arrives at a male's tunnel, he stops his loud chirping but continues to court her with a third and softer song that cannot be heard at a distance, thus tending to avert unwelcome intrusions by other males.

If we were to look closely, we might find one or two silent male crickets lurking in the vicinity of a calling male. These males do not sing, although they do have the chirping apparatus. Known as satellite males, they seek mates in a different way. When a female approaches a singing male, the satellite male silently sneaks out to intercept her and is sometimes successful in seducing her. Can there possibly be an advantage to being a silent satellite male rather than a more direct and forceful caller?

A probable advantage becomes obvious if we consider the cricket in its ecological context, as an organism that lives in an ecosystem with other organisms. Crickets are parasitized by one of these other organisms, a fly that belongs to a large family of parasites, the tachinids. The female tachinids are attracted to male crickets by their calling songs. (They can be lured to a loudspeaker that plays the call of a cricket.) The flies lay their eggs on the male cricket, and the maggots that hatch from the eggs burrow into the body of the cricket and ultimately kill it. Thus, in a year when the parasitic flies are abundant, the timid satellite males may actually father more progeny than the more romantically insistent, calling males, most of which are killed by parasites.

Anyone who has experienced an emergence of periodical cicadas, sometimes loosely called locusts, knows that these insects, although virtually harmless, are hard to ignore. There are countless millions of them, and they make an almost deafening uproar. The emergence of Brood XIII in the northern suburbs of Chicago in 1990 was a media event. Cicadas were on everyone's mind. The *Chicago Tribune* published recipes for cooking cicadas; bars held cicada-eating contests; and a woman from Winnetka sold jewelry made from cicadas. To most suburbanites the sounds of these insects were a distracting cacophony, but to the cicadas they were urgently important messages concerning reproduction and the integrity of the species.

The nymphs, or immature forms, of this brood of cicadas hatched from eggs in 1973. They spent the next seventeen years in the soil sucking sap from the roots of trees. In the spring of 1990 they were full grown and dug their way to the surface to metamorphose to the adult stage as they clung to the trunks of trees or other surfaces that gave them purchase, perhaps the wall of a house. The empty, molted skins, looking like ghosts of the nymphs, were everywhere. The females laid their eggs in punctures that they made in twigs. Six or seven weeks later the nymphs hatched, dropped to the ground, and disappeared into the soil. They will not be seen again until the year 2007.

The male cicada's sound-producing organs are in two large air-filled cavities on his underside near the base of the abdomen. These organs are anatomically complex, more so than the larynx of a human or the syrinx of a bird, but their core is a springy,

domelike timbal. A pull of the muscles attached to the timbal causes it to buckle inward and make a click—like pressing down on the top of a can. When the muscles relax, the elastic timbal springs back outward, making another click. Cicadas produce about 390 clicks per second, making what sounds like a high-pitched squawk to our ears. To another cicada it may sound like the epitome of musical perfection. Female cicadas are mute, but they, like the males, have ears located on the underside near the base of the abdomen.

Richard D. Alexander and Thomas E. Moore of the University of Michigan found that male periodical cicadas produce three sound signals, each of which serves a different function. Individuals that are startled into flight make a disturbance squawk. This sound can also be evoked by grasping a male in the fingers. The calling song and the courtship song are both used in bringing the sexes together. Both males and females respond to the calling song by joining the singer in the tree from which he calls. Over a period of about two weeks, hundreds or even thousands of males congregate in the same tree and sing as a perfectly synchronized chorus. Females do not congregate; they mate quickly and then leave to lay their eggs. Large and loud aggregations seem to be particularly attractive to females, and males that join a chorus are more successful at finding mates than are lone males. The final song, the courtship song, is sung by the male as he approaches a female that has been attracted to his "chorus tree," but vision seems to be as important as hearing in close-range courtship.

Although periodical cicadas emerge in tight synchrony, a brood actually consists of three different species that generally maintain their integrity; in other words, they do not hybridize with each other because each species sings a different calling song. You will probably be able to pick out these three distinct songs from the general uproar if you listen carefully throughout a day. You will also discover that each chorus includes the males of only one species. Only one calling song comes from each "chorus tree." Not only do these songs sound different from each other, but mismatched matings between species are made even less likely because they all sing at different times of the day. One species usually sings in the morning, another around the middle of the day, and the third in the afternoon. It did not come as a surprise to biologists that these species sing different calling songs and thus generally avoid mating with members of the other closely related species. A species remains a separate entity because it is reproductively isolated from other species.

Many insects initiate courtship in response to visual stimuli rather than scents or sounds. For example, the males of many species of flies, including certain parasitic flies, flesh flies, and flower flies, watch for females from a perch: a leaf, a branch, or even a rock that gives them a good view of the surrounding air space. If an insect of about the right size flies into view, the male darts out to grapple with it. If he contacts a female of his own species, the pair will hide in nearby foliage and copulate. If the intercepted insect is a male of the same species or a member of some other species, the contact ends quickly and the startled object of this unwelcome attention makes a hurried retreat. This seemingly aggressive encounter may be only the consequence of the male's failure to identify the intruder without making contact to taste a pheromone. It may also be interpreted as the territorial defense that is practiced by many mate-seeking male insects. The male apparently recognizes females by means of a chemical signal that he perceives only at close range, either a smell or a taste. Butterflies have organs of smell on their antennae, but both flies and butterflies have organs of taste on their feet and could sample the flavor of insects that they contact. It is easy to prove to yourself that a perched, mate-seeking male cannot recognize a female by visual stimuli alone. All that you have to do is to flip in front of him some small object, a bit of bark or twig, of about his own size. He will dart out and briefly grapple with it.

In 1942, Niko Tinbergen, the Dutch biologist who in 1973 shared a Nobel prize with Konrad Lorenz and Karl von Frisch for his work in ethology, the biological approach to the study of behavior, demonstrated that a European butterfly, the drably colored grayling, can locate females of its own species by sight, but that visual signals alone do not induce copulation. Male graylings will pursue various insects, and they can even be induced to chase paper dummies. But on close contact they reject all but female graylings.

Tinbergen and his coworkers attached paper dummies of various sizes and shapes to pieces of fishing line tied to rods and waved them in front of sitting male graylings. Dummies of many sizes, colors, and shapes were pursued by males, some that actually resembled graylings and some that were mere rectangles or circles of paper. The most attractive dummy was an object, not necessarily shaped like a female, that was brought as close to the male as possible, that was as dark as possible, and that was moved in a fluttery, butterflylike manner.

When male graylings approach females of their own species, they presumably recognize them at close range by means of a pheromone. If a male convinces a receptive female to land, he lands near her and induces her to mate by means of an elaborate courtship display that culminates with his making a graceful bow to the female as he catches the tips of her antennae between his front wings. The upper surface of each of the male's front wings bears a patch of specialized, pheromone-producing scales that rub against the female's antennae during the course of the bow and that may act as an aphrodisiac.

Fifteen years later, a German ethologist, Dietrich Magnus, extended Tinbergen's experimental approach to another and more colorful butterfly, the silver-washed fritillary. Using motionless pieces of colored paper, he found that sexually motivated males, easily recognized by their searching, zigzag flight, are attracted to orange, the color of the female. But, as with many other butterflies, both females and males that are not searching for mates are attracted to green, the color of the leaves on which they rest, or to blue or yellow, the colors of the blossoms from which they sip nectar.

Since males usually encounter flying rather than stationary females, Magnus attached dummy butterflies made of paper to the ends of two six-foot-long arms protruding from a motor-driven "merry-go-round." When the merry-go-round rotated, the dummies appeared to fly. Males responded to only three characteristics of a dummy: its size, its color, and the frequency with which it appeared to flicker—as the wings flapped or, on a less realistic model, as a cylinder with alternating black and orange bands rotated. Shape did not matter. Males were as strongly attracted to circular or triangular pieces of paper as they were to butterfly-shaped dummies. Orange dummies with simple black markings were more attractive than dummies covered with a veneer of the more complexly patterned real wings. Male fritillaries were more attracted to "supernormal" dummies than to realistic dummies. Just as birds can be tricked into abandoning their own eggs to incubate a supernormal or larger-than-life artificial egg, male fritillaries preferred dummies that were more orange and four times as large as a real female and that appeared to flap much more rapidly than a real female. The supernormal dummy is the vice, the *Playboy* centerfold, of the male silver-washed fritillary.

Although male fritillaries eagerly pursued paper dummies, they never attempted to court them or to copulate with them. They lost interest when they got within four inches. But Magnus found that they did not lose interest if the visual stimulation of a dummy was accompanied by the scent of a female. If a live female in a small cage was hidden under a stationary dummy, so that only her scent was perceptible, approaching males went on to court the paper dummy and even tried to copulate with it. They never left semen on a dummy, although they were obviously aroused. Additional stimulation, probably by touch, is required for completion of the sexual act.

In his text on entomology, first published in 1920, John Henry Comstock of Cornell University, the first great teacher of entomology in the United States, wrote the following words:

> During some warm, moist evening early in our northern June we . . . see here and there a tiny meteor shoot out of the darkness near at hand, and we suddenly realize that summer is close upon us, heralded by her mysterious messengers, the fireflies. A week or two later these little torch-bearers appear in full force, and the gloom that overhangs marshes and wet

meadows, the dusk that shrouds the banks of streams and ponds, the darkness that haunts the borders of forests, are illumined with myriads of flashes as these silent winged hosts move hither and thither under the cover of the night.

While flies, butterflies, and cecropia moths make little use of visual characteristics to *identify* potential mates, the fireflies, or lightningbugs (which are actually beetles), use flashing, luminescent signals both to locate and to recognize their mates, but it is likely that they also make some use of pheromones, especially at close range. A diverse array of organisms are bioluminescent: they can create "cold light" by biochemical means. Among the insects, these include beetles other than the fireflies, certain true bugs, and the maggots of some flies. (Most people, including many entomologists, use "bug" to refer to any insect, but technically speaking, this word applies only to members of the Order Hemiptera.) But insects are not alone in being bioluminescent. Certain jellyfish, squids, deep-sea fish, and even single-celled organisms may be bioluminescent. The infamous red tides of the oceans, which can kill fish by the millions and cause clams and mussels to be toxic to humans, consist of countless billions of tiny unicellular protozoa, distant relatives of the amoeba, that bioluminesce when they are disturbed. At night boats in affected waters leave a glowing wake and dipping an oar into the water brings forth a bright flash of light.

The fireflies are by far the most familiar of the bioluminescent organisms. On summer nights just after dusk, including those agreeable June evenings when male cecropias are on the wing, we see the brief sparkle of fireflies as they fly over our lawns or rest on the grass or the other plants in our gardens. To fireflies, these flashes of light are signals of life-or-death importance. They herald the birth of progeny and continuing life to those that are successful in finding a mate. To a few unlucky males, they foretell death in the jaws of a predator.

First, the lucky males. Mate-seeking males of the common species in eastern North America emit brief coded flashes that, owing to their number, their duration, and the intervals between them, are characteristic of the signal sender's sex and species. To avoid confusion, these signals need to be specific, because several species of fireflies may be seeking mates in the same place at the same time. The females, perched on some slight rise such as the tip of a grass blade, flash only in response to a male of their own species—with the one exception discussed below. The flashes of a male and a female of the same species may be quite different, but both are characteristic of the species and each is characteristic of its sex. Once a male and a female are aware of each other's signals, they exchange flashes until the male finally locates the female and copulates with her.

Now for the unlucky males. Males of the *Photinus* group of fireflies, the ones that we have been considering thus far, are sometimes eaten by the females of the *Photuris* group of fireflies. These carnivorous females respond to courting males of their own species with their own species-specific flash signal. They mate with their own males, and they do not usually eat them. But their response to the signals of other firefly males is different. They respond to the flashes of *Photinus* males with the falsified flash signal of the *Photinus* female. The carnivorous *Photuris* females have broken the *Photinus* code. These firefly *femmes fatales*, as they were dubbed by James E. Lloyd, the discoverer of this type of aggressive mimicry, devour the hapless *Photinus* males that they lure in by deception.

Fireflies and other bioluminescent insects have captured the attention of people wherever they occur. The British entomologists William Kirby and William Spence, in the 1846 edition of their *Introduction to Entomology*, first published in 1815, describe the use of living fireflies as ornaments: "In India, as I am informed by Major Moore and Captain Green, [the ladies] even have recourse to fire-flies, which they enclose in gauze and use as ornaments for their hair when they take their evening walks." In his 1865 *Curious Facts in the History of Insects*, Frank Cowan reported that "the poorer classes of Cuba and the other West India Islands make use of these luminous insects for lights in their houses.

Twenty or thirty of them put into a small wicker cage . . . will produce quite a brilliant light."

Like fireflies, crickets, and cecropia moths, hundreds of thousands of other insects locate mates by using one of the three senses that can receive signals from a distance: vision, hearing, or smell. But another way to find mates is to do as people do, to look for them in places where members of the opposite sex are likely to be in the course of their daily lives. In our complex human society, there are many such meeting places: churches, supermarkets, laundromats, and libraries, to mention just a few. But insects lead far simpler lives. The females, totally preoccupied by the drive to reproduce, are concerned only with obtaining sperm to fertilize their eggs, with taking in enough nourishment to keep themselves alive and develop their eggs, and, in most species, with laying their eggs in suitable places where their offspring will find the food and other resources that they require. Thus for most male insects—almost always the sex that searches for a mate—there are just two likely trysting places: where the females eat or where they lay their eggs.

Among the insects that rendezvous at feeding sites are solitary bees of the genus *Hoplitis*. The females, and the males as well, eat the nectar and pollen of only one kind of plant, the viper's bugloss. Males seek out clumps of these plants and remain by them for long periods of time as they fly from blossom to blossom searching for sexually receptive females that have come in to feed.

Tsetses, flies that in Africa transmit the protozoan that causes sleeping sickness in humans, eat blood—occasionally the blood of humans but more often the blood of large grazing mammals such as wildebeests and zebras. Males follow or perch on these animals and try to mate with females that arrive to take a blood meal. In an experiment, male tsetses behaved similarly toward a vehicle that was draped with a gray blanket and driven slowly across the savannah. They followed the vehicle as it moved but alighted on the blanket when the vehicle stopped. It appears that male tsetses rely mainly on dimly perceived visual cues to identify their mobile trysting sites.

Other insects rendezvous where the females lay their eggs. Walnut husk fly maggots eat only the tissue within the outer, fleshy husks of walnuts, usually the native black walnut in eastern North America, and that is, of course, where the females lay their eggs. Males defend individual walnuts against other males and mate with any females that come in to lay eggs. All damselflies lay their eggs under water on aquatic plants, where their predaceous offspring are likely to find other aquatic insects to eat. Egg-laying females fly to a plant that protrudes above the surface and then crawl down into the water or just reach down with their long abdomens. Male damselflies stake out territories that include suitable egg-laying sites and defend them against other males. Exceptionally enterprising males may assemble a small harem of females.

Many male flower flies seek females both where they feed and where they lay their eggs. One of my graduate students, Chris Maier, found that in central Illinois in June, males of a large, hairy flower fly that resembles a bumble bee arrive early in the morning at blossoming wild roses, elderberries, or dogwoods that grow at or near the edges of woodlands. They spend most of their time making patrolling flights around these bushes, paying particular attention to blossoms. If they spot a female feeding from a blossom, they pounce on her and almost immediately carry her off into the foliage where they remain hidden as they mate. Except for a few stragglers, the males leave the blossoming bushes at about eleven in the morning, when the hottest part of the day begins. During the hot afternoon, males can be found in shaded woodlands at the moist rot cavities in oaks in which the females lay their eggs. They defend territories surrounding these treeholes against other males and pounce on any females that arrive to lay eggs. The females always seem to be willing to copulate.

Some mammals, birds, and insects find mates by congregating at specific assembly areas called leks, a behavior reminiscent of the groups of eager people in singles bars and the gatherings of college students on the beaches of southern Florida and Texas during the spring break from classes. Unlike these "leks" used by humans, those used by most other species offer neither food

nor drink to the congregants—not even places to lay eggs. The only resources present are members of the opposite sex. Every spring the sage grouse of our western plains gather at the same leks that they have used for decades or perhaps for much longer. The elaborately feathered males perform eye-catching displays on small territories within the boundaries of the lek. The smaller and plainer females choose among the many males on the lek, usually preferring high-status males that have secured and defended against other males a desirable territory near the center of the lek.

There have been many reports of insects forming leks, but one of the first was by a zoologist named John A. Chapman. On a warm day in July of 1951, Chapman climbed isolated Squaw Peak in western Montana. On the barren summit he found many flies of several species zooming about or perching on the bare rocks. There were no discernible resources for these insects on the summit—no water, food, or suitable places to lay eggs. Observations by Chapman and many others have since shown that numerous species of flies, beetles, wasps, bees, and butterflies rendezvous at similar sites for the purpose of locating mates. Males usually outnumber females at these leks, because males remain at the lek to mate repeatedly while females leave immediately after copulating. It may be that sage grouse use the same lek year after year because each generation learns its location from members of a previous generation. But this is not likely in insects, because the adults of succeeding generations are usually separated from each other by weeks, months, or even a year or more. Lek-finding information is presumably transmitted genetically from generation to generation—certainly not the precise location of a lek, but rather a program of behavior that will lead an individual to a likely location. Mating aggregations will inevitably form if individuals are genetically programmed to visit mountaintops, hills, or other prominent topographic features of the landscape, the very places where insect leks are usually found.

Like cecropias, some insects and other animals copulate immediately after they first meet—abruptly, unceremoniously, and with no apparent attempt at courtship or foreplay. But other animals, including many insects, fish, birds, and mammals, do not copulate until after they have gone through a period of courtship. Animal courtships are stereotyped, ritualistic, and uniquely characteristic of a species. Courtship rituals may be brief or lengthy. They may be relatively simple, but they are often elaborate.

The peacock's fabulous tail, so bulky that it is a hindrance in everyday life, is useful only in his spectacular courtship of the peahen. Male sage grouse court females by strutting and posturing before them, fluffing up their feathers, fanning out their tails, and inflating two huge and conspicuous yellow air sacs on either side of the throat. The male satin bowerbird of eastern Australia defends a territory in which he builds an elaborate theater with a stage from which he courts the female. He erects two parallel walls of interlacing twigs stuck into the ground, thus forming a bower large enough to accommodate a female. He paints the inner walls of the bower with the crushed pulp of blue berries. (The bower is used only in courtship. The female builds a separate nest in a tree.) In front of the bower he makes a large platform or stage of twigs and grass that he decorates with many small objects, mostly blue, that he collects here and there: feathers, flowers, berries, and even shards of blue glass and scraps of blue paper. Females watch from behind or within the bower as the male displays, ruffling his feathers and fanning his tail, all the while holding a blue object in his bill.

The courtship of the queen butterfly, a species from the southern United States closely related to the familiar monarch, was studied by Lincoln Brower, who was then a professor at Amherst College. Males of the species make cruising flights in search of females. When a male sees a flying female he rapidly overtakes her in straight-line flight. The manner of his flight suddenly changes to a rapid up and down bobbing when he is a few inches directly above her. At this point he probably recognizes her as a female of his own species by her scent. The male now does something charmingly graceful. As he flies above the female, he extrudes from the end of his abdomen two normally hidden organs called hairpencils. They are not organs of copulation; rather, they shower the antennae of the female with tiny, glisten-

A male queen butterfly flies above a female as his hairpencils shower her antennae with an aphrodisiac dust

ing, dustlike particles that contain an aphrodisiac pheromone. When the hairpencils are extruded, a tuft of delicate hairs at the tip of each one fans out into the shape of a miniature brush. It is from these hairs that the male's seductive dust falls onto the antennae of the female.

A responsive queen then alights on a plant. The male hovers over her and once again dusts her antennae with the aphrodisiac dust. If she is willing to copulate, she brings her wings together over her back. The male then lands beside her, wings folded over his back, and contacts her genitalia with his own. Shortly after the pair couples, the male opens his wings and snaps them shut, perhaps in exuberance or as a signal to the female of what is to come next. In any case, wing-snapping is followed by a post-nuptial flight. The male flies off with the passive female dangling from his genitalia, held tight by his genital claspers. She makes no movements but keeps her wings folded together and her legs tucked up against her body. The flight is usually short, and the couple quickly settles in the vegetation. After a brief time, the male passes his sperm into the female, but the two may remain coupled for several hours more.

Meetings between the sexes often involve an element of conflict. Both individuals may be ambivalent; the desire to flee may accompany the desire to proceed. A territorial male may not know whether an approaching individual is a female or another male that will try to displace him. In predatory species, a male approaching a female hopes that she will accept him as a mate but may also fear that she will eat him. Courtship rituals mollify such conflicts.

Web-spinning spiders are a case in point. When a hungry female feels her web twitch and vibrate, she hurls herself across the web and quickly injects venom to paralyze the entrapped prey before it can break free. A courting male, who must approach the female by climbing over her web, is accordingly in great danger of being killed and eaten before he can so much as touch the female. In order to avoid this fate, the male sends a telegraphic message to the female by plucking a strand of her web in a code that she will understand, thus informing her that he is a suitor rather than a meal. He then approaches cautiously and further pacifies the female by gently stroking her body. Some males are, nevertheless, killed by females. They are often devoured just after insemination if they do not hurry away.

In one of Gary Larson's "Far Side" cartoons, two matronly praying mantises are having an argument. One says to the other, "I don't know what you're insinuating, Jane, but I haven't seen your Harold all day—besides, surely you know I would only devour my *own* husband!" Female mantises do not always eat their mates, but when they do they begin by eating their heads. Amazingly, the male will initiate copulation even after his head and brain are gone, and he will ultimately inseminate his mate after she has eaten even more of his body. By so doing the doomed male salvages what he can. Being eaten by his mate is not a total loss. He becomes a meal that will help her to survive until she has laid the eggs that will become his sons and daughters. He has at least inseminated this one female, but he will never inseminate another. Other carnivorous insects have evolved courtship rituals that allow males to survive copulation so that they can inseminate several females.

Males of the predaceous hangingflies, a subgroup of the scorpionflies, divert the gastronomic attentions of their carnivorous females away from themselves by presenting them with another insect to eat. A male that intends to seek a mate first hunts for insects. If he catches a small one, he will probably eat it himself. If he catches a large one, he saves it until he attracts a mate. With the freshly killed insect held in his hind legs, he hangs from a plant by his front legs as he releases a sex attractant pheromone. When a female arrives, he presents her with the insect and copulates with her as she eats.

The size of the prey insect determines how long copulation lasts. If it is large, the female may feed and allow the male to copulate for as long as twenty minutes, ample time to transfer his sperm into her body. If the prey insect is small, she may reject him before copulation can begin, or she may eat and copulate with him for less than five minutes, not enough time to transfer his sperm to her. Female hangingflies know a good thing when they see one. They mate frequently and obtain virtually all of their food from males. By feeding the female, the male not only ensures his own survival but also enhances his own fitness by contributing nutrients that enhance her ability to develop and lay the eggs that his sperm will fertilize.

The dance flies, true flies that use their beaks to suck juices from their insect prey, are a modestly numerous family of about 2,000 different kinds. (The true flies are the two-winged members of the Order Diptera, as contrasted with dragonflies, mayflies, and other four-winged insects that are not true flies.) The males of many species present their mates with prey. But in other dance flies the presentation of prey to the female has evolved into a ritualistic and purely symbolic gesture. Different degrees of this behavior, which may represent evolutionary steps leading to the ritualization of the presentation of prey, can be seen among the North American species. The males of some dance flies do not present the female with a meal during courtship, and they are occasionally eaten by their mates. In other species, the male avoids being cannibalized by presenting his mate with a nuptial gift, a freshly killed insect that may be as large as he is. Copulation lasts only as long as it takes her to eat the insect. In another dance fly we see the first step toward ritualization; the male tangles a few silken threads about the nuptial gift. In other species, the male spins a silken balloon that he attaches to a large and freshly-killed insect that the female will eat. In yet another stage of the ritualization, the male presents the female with a much larger balloon to which is attached a minute and often dried-out insect that she does not even attempt to eat. In the putative final stage, the males of some species offer the female only a large empty balloon that holds no prey at all. She does not eat or otherwise use the balloon, but she does not eat the male. The balloon has become the symbol of the prey. Why the females are willing to accept nothing more than a symbol of a meal remains a mystery.

Courtship rituals have another and more universal function than placating carnivorous mates. Through courtship, males and females get to know each other before they make the most momentous decision of their lives, choosing the mother or the father of their offspring. Females judge courting males by their appearance, by how well they perform the courtship ceremony and, in territorial species, by the quality of the territory that the male has acquired. By choosing high-quality males a female increases the probability that her progeny will survive to reproduce, thus enhancing her own fitness. Clean, comely males are likely to be healthy and will probably father healthy children. Sexy males that perform the courtship ritual with precision and panache are likely to father sexy sons, Don Juans who will mate with many females and pass their mother's genes on to many grandchildren.

The males and females of many species of birds and mammals and a few species of insects share parental care. But a male that does not help with parental care may invest no more than a few drops of semen in his offspring. Thus he can profit by scattering his seed among many females, gambling that some of them will be good mothers. But, as we have already seen, females are generally less promiscuous than males and more selective in choosing mates. They cannot afford to be otherwise because they invest so much more in reproduction than do the males. After a sage hen mates she goes off alone and spends nearly two months incubating eggs and caring for chicks. Although a cecropia female mates with the first male to reach her, there is, nevertheless, an element of selectivity involved. The first male won a race with other males and is thus likely to be superior to them in alertness, fleetness, and the sensitivity of his "nose."

There are exceptions that prove the rule: a few species in which the sex roles are reversed and the males make a greater investment in their offspring than do the females. In these species, the males tend to be the coy ones and are more selective and less promiscuous than the females. In these animals, courtship roles are essentially reversed. After mating, the females of certain giant water bugs glue their eggs to their mate's back and go off to seduce more males while the father gives the eggs elaborate care. He does not mate again as long as they are on his back. Sex roles are even more completely reversed in those charming little marine fish known as seahorses. The "liberated" female uses an intromittent organ to pump her eggs into a pouch in the belly of the male, where he fertilizes them. The male broods the eggs; and in some species he even has a "placenta" that nourishes the embryos; and ultimately he "gives birth" to young that are miniature versions of himself or his mate.

In the vast majority of insects and other animals it is the females that are the more selective and the less promiscuous. Thus willing females tend to be in short supply, and in many species there are not enough opportunities to satisfy the appetite for mating of all of the males in a population. Hence, as we have seen throughout this chapter, males are in competition for the favors of females.

After the Courtship's Over

Successful courtships end in insemination. In the vast majority of insects, insemination is accomplished through copulation and is thus more or less comprehensible from our own point of view. The male cecropia, for example, uses his intromittent organ, which we can call a penis by analogy, to inject sperm into the genitalia of the female. But this is not always the way it works. A few very primitive insects and some other arthropods (jointed-legged relatives of the insects) accomplish internal fertilization indirectly; the male "casts his seed upon the ground" and leaves it to the female herself to place it within her body. Spiders, eight-legged relatives of the six-legged insects, did not evolve a penis but instead use modified legs to place sperm in the body of the female. A male bed bug uses his intromittent organ to pierce the wall of the female's abdomen and inject sperm into her blood rather than her genitalia—a behavior aptly known as traumatic insemination. For reasons that we will consider later, damselflies and dragonflies evolved a "secondary penis" to substitute for the original. These and various other methods of fertilization used by insects and their relatives, raw material for a *Kama Sutra* for arthropods, are bizarre from the human point of view, but they do make good evolutionary sense when they are examined more closely and in context.

The remarkable evolutionary success of the insects would have been impossible if their ancestors had not made the transition from living in the seas to living on dry land, and this transition would not have been possible if those ancestors had not evolved the capacity for internal fertilization. Sea anemones, jellyfish, starfish, and many other animals are bound to their life in the sea because, like American oysters, they practice external fertilization; their sperm and eggs are simply released into the water, where they must find each other. A few animals that have not evolved internal fertilization, notably frogs and toads, have made partial transitions to the land but must return to the water to reproduce. A male bullfrog mounts his mate and clasps her fiercely, but he can only release his sperm into the water as she releases her eggs. But insects have freed themselves from the water by their capacity for internal fertilization. Whether by copulation or indirect means, their sperm end up in the moist environment of the female's body, thus protecting these tiny bits of life from a dry environment in which they would surely desiccate and perish.

From a human perspective, indirect internal fertilization, such as practiced by the springtails, scorpions, and silverfish, seems clumsy and oblique. Nevertheless, internal fertilization of any sort, no matter how awkwardly achieved, is vastly more efficient than external fertilization as practiced by the distant, seagoing ancestors of the arthropods, possibly wormlike creatures similar to certain segmented, marine worms that still practice external fertilization. The fantastic palolo worms, for example, develop eggs and sperm in the rear half of their long, many-segmented bodies. At a time determined by the phase of the moon, the headless posterior halves of the worms break off from the rest of the body and squirm to the surface of the sea, where eggs and sperm are shed into the water. This is, of course, extremely wasteful, but the probability that a sperm or an egg will go to waste is at least slightly decreased because all of the worms in an area release their eggs and sperm at the same time. Off the shores of Bermuda, palolos swarm exactly three days after the full moon and, it is claimed, usually at precisely fifty-four minutes after sunset.

If we scan the various ways in which arthropods go about fertilization, we see a variety of behaviors that probably represent evolutionary steps leading from primitive external fertilization, even as practiced by such nonarthropods as the palolo and other marine worms, to internal fertilization accomplished by the more sophisticated techniques of copulation. It is as if we can see nature, via natural selection, groping for an efficient method of fertilization. Since arthropods that copulate have diversified to produce about 75 percent of the known species of animals, it seems fair to conclude that copulation is an evolutionarily sound solution to the problem of efficient fertilization. Further testimony to the usefulness of copulation is that the reptiles, birds, and mammals evolved this behavior quite independently many millions of years after the insects did. In his *Sexual Selection and Animal Genitalia*, William G. Eberhard discusses many interesting aspects of internal fertilization.

To this day, a few very primitive arthropods—namely the horseshoe crabs, also known as king crabs—are capable of only external fertilization. These peculiar creatures, not really crabs at all, but distant relatives of scorpions and spiders, are living fossils whose history goes back 350 million years. On summer nights, horseshoe crabs, which may be as much as eighteen inches long, swarm from the shallows up onto the beaches of the east coast of North America and the east coast of Asia to reproduce. The female scoops out a shallow, water-filled depression in the sand in the zone between the high and low tide lines and then lays her eggs as a male mounted on her back spills his sperm into the nest. This is doubtlessly an improvement, albeit a modest one, over simply releasing eggs and sperm into the sea. The male does, after all, seek out the female and, like a frog or toad, deposits his sperm directly upon her eggs.

The symphylans, primitive, land-dwelling arthropods that look like tiny centipedes, also practice external fertilization—but with a twist that makes their method more efficient than that of the horseshoe crabs. Males deposit drops of semen on the ground whether or not females are present, each droplet held above the surface of the soil by a tiny stalk. Fertile females search for these droplets of semen, bite them off their stalks, and collect them in special pouches associated with their mouthparts. When the female lays an egg, she picks it up with her mouthparts and, before depositing it, "mouths" it as she smears it with tiny droplets of semen.

Springtails, which are among the arthropods that accomplish internal fertilization without copulation, were once thought to be primitive insects because they have six legs, but they are now considered to be a separate group of arthropods. Few of us have seen these tiny creatures, but all of us have passed by or sat over many springtails. A square yard of soil generally contains thousands of them. Some springtails have elaborate courtships, but others do not even bother with courtship. The males simply deposit on the ground numerous tiny droplets of semen attached to stalks, even when there are no females around. Females that are ready to reproduce search for these droplets of semen and, without any assistance from a male, take them into their genital opening.

The courtship of a European scorpion, an eight-legged arthropod related to spiders and ticks, was elegantly recounted by J. Henri Fabre, the nineteenth-century "Homer of the insects," in his *Souvenirs entomologiques*, recognized as a milestone in the study of insect behavior as well as a contribution to French belles lettres (see the English version by Edwin Way Teale). The small male scorpion, often ultimately eaten by his larger inamorata, grasps the female's pincers with his pincers and walks backwards as he leads her in a macabre "*promenade à deux.*" As they dance, the male stings her several times with the fang-like stinger at the end of his abdomen. But she is immune to his venom. Perhaps these stings function like love bites. The male deposits a packet of sperm on the ground at the end of this dance and carefully positions the female over it. She then opens the packet, pinches it so as to force the sperm into her genital cavity and, finally, eats the empty packet.

The little silverfish, often unnoticed and essentially harmless guests in our homes, especially city apartments, are true insects but are so so primitive that they never evolved wings and still practice indirect internal fertilization. As reported by H. Sturm, in a profusely illustrated article in German with an English summary, an elaborate courtship precedes fertilization. First, the male and the female stand face to face. With their heads almost touching, they alternately quiver their antennae and swing their heads from side to side. This is followed by something like a game of tag. One partner turns away from the other, runs a short distance, and then returns to touch the other. One of the functions of this courtship is to bring both partners to the same pitch of excitement, or readiness for fertilization, at the same time. When this has been achieved, the male spins several silken threads across the female's path. One end of each thread is attached to the ground and the other is attached higher up on some vertical surface. He then places a packet of sperm on the ground directly beneath these threads. When the running female bumps into these threads, she stops, carefully searches for the packet of sperm, and takes it into her genitalia.

A strange and altogether unique form of "copulation" is practiced by spiders, as recounted by W. S. Bristowe. As I have already pointed out, male spiders use a pair of modified legs as substitutes for the penis that they never evolved. Spiders have four pairs of walking legs, but in both sexes a modified fifth pair, not used for walking, is situated on either side of the mouth. Known as pedipalps (literally "feet for feeling"), they have been aptly described as the "hands of the spider." In mature males, the tip of each pedipalp is modified to form an intromittent organ that works like a rubber-bulb syringe. Before beginning to court a female, the male loads his pedipalps with semen. First he spins a small silken web on which he deposits liquid semen from the genital opening near the base of his abdomen. Then he sucks the semen into the syringes at the tips of his pedipalps. After a successful courtship, he thrusts one of these "hands" into one of the genital openings of the female and injects his semen.

In all insects other than the silverfish and their few primitive, wingless relatives, insemination is accomplished by copulation. That is, the male uses his penis to place his sperm directly within the body of the female. The mayflies, certainly the most primitive of the winged insects, differ from other insects in that the females have two genital openings and the correspondingly well-equipped males have two penises. In all other insects, with only one minor exception, the females have only one genital opening and the males have only one penis. The genital apparatus of a male insect is generally far more elaborate than that of most male mammals. In many insects the penis is flanked by intricate accessory genital structures, notably claspers that grip the genitalia of the female. The penis itself may be short or as long as the rest of the body; it may bear complex hooks, spines, or inflatable sacs. As we will see, at least some of these complexities are adaptive features that serve to enhance the fitness of the male. They may permit him to interfere with the sperm of a preceding male. Or they may enable him to retain his hold on a female when other males try to displace him and take over.

The genital apparatus of a male insect is characteristic of his species. It usually differs greatly between species but varies only slightly among males of the same species. It was once suggested that the genitalia of males fit those of females like a key fits a lock, and that this lock and key arrangement prevents mating with a member of the wrong species. This may occasionally be the case, but we now know that this is not a universal reason why so many insect species differ from all others in the structure of the male genitalia. As I will discuss below, the specialized penises of males sometimes enhance their ability to compete for females with other males of their own species.

In most insects, the fertilization of an egg proceeds in a more complex fashion than in humans. This complexity gives the female more discretion in the process of fertilization and makes possible a more economical use of sperm. A human female receives about 70 million sperm at each copulation. Some of them die in the relatively inhospitable environment of the vagina, but others move into the uterus and up the fallopian tubes, where only one of this great swarm of sperm fertilizes the egg as it moves down from the ovary.

Unlike humans, almost all female insects, except for the primitive mayflies, have an internal storage pouch for sperm, the spermatheca, that branches from the insects' analogue of a vagina. The sperm are deposited in this pouch during copulation or are deposited in the vagina and then move into the pouch shortly after copulation. In this receptacle, the sperm can be kept alive and healthy for days, weeks, months, or even years in some insects. It is the sperm pouch that makes it possible for a female cecropia to mate only once but to continue laying fertile eggs for several days. As do other insects, she fertilizes each egg that she lays by releasing just a few sperm as the emerging egg passes the opening to the storage pouch. In some insect species, the female need release only one sperm for each egg that is to be fertilized.

Only a few vertebrates, not including humans, have the ability to store sperm for long periods of time. Sperm rarely survive for as long as 7 days in the reproductive tract of human females. Although bats do not have a specialized receptacle for the storage of sperm, they can, depending upon the species, store them in the reproductive tract for from 16 to 200 days. Some reptiles, lizards, snakes, and turtles do have a receptacle for sperm. Turtles can store viable sperm for as long as four years, and snakes can store them for as long as five years and perhaps even for seven years. In insects and vertebrates alike, the ability to store sperm makes it possible to mate at the most propitious time for that behavior and to lay eggs or implant a fetus at the most favorable time for reproduction.

The usefulness of the sperm pouch can be seen to good advantage in the honey bee. Among the 30,000 or more females in a hive, all but the one and only queen are nonreproducing female workers. Only the queen is capable of laying eggs, but she never mates in the hive with her several hundred resident sons, the drones. A young virgin queen's only opportunity to mate comes when she leaves the colony to fly to a site where queens and drones from different colonies congregate for the purpose of mating. There she mates on the wing with several drones and returns to the colony with their sperm, usually about 6 million of them, stored in her sperm pouch. Since she will never mate again, these sperm must suffice for the duration of her reproductive life. During her tenure in the colony, usually about two years

but as long as eight years if she is coddled by the beekeeper, she lays as many as 2 million eggs, less than 100,000 destined to become drones and all the rest destined to become females. The queen bee can ration out only 6 million sperm to fertilize about 2 million eggs. She can at will predestine an egg to become either a male or a female. Eggs fertilized by a sperm become females and unfertilized eggs become drones. In ants, bees, wasps, and their close relatives unfertilized eggs become males. In almost all other insects, eggs destined to become either sex must be fertilized by sperm.

The ability of female insects to store sperm for long periods presents an opportunity for males to enhance their fitness by monopolizing the fertilization of some or all of a female's lifetime production of eggs. A male generally provides a female with an overabundance of sperm, and even if she continues to lay eggs long after mating, her storage pouch will contain enough to fertilize them all. But in most insect species there is a complication. If the female mates with another male before laying any eggs, the first male's sperm may fertilize only a few of her eggs or often none of them at all. The reason for this is a phenomenon that entomologists call "sperm precedence." The last sperm to enter the female's storage pouch take precedence over the first to enter. Thus the last male to mate with a female may fertilize most or all of her eggs. Consequently, some male insects have evolved means by which they try to keep their mates chaste, to prevent them from mating with other males. But they have also evolved equally clever ways to cuckold other males—sometimes with the willing cooperation of the female—by circumventing these chastity-enforcing mechanisms.

The females of some insects may, after only one successful insemination, reject the sexual advances of all males. Cecropia females, for example, are among the insects that are sexually turned off for the rest of their lives after copulating only once. Females of many other species are turned off only temporarily, perhaps for a matter of hours or days. In either case, the first male is likely to benefit, because his mate's permanent or temporary abstinence ensures that all or at least some of her eggs will be fertilized by his sperm rather than by those of some later-arriving male. Mating several times, as do queen honey bees and many other insects, may be to a female's benefit even if the first male provides more than enough sperm to fertilize all of her eggs. Why trust all of her eggs to only one male? Using the sperm of several fathers will produce genetically more varied offspring, and the offspring of some fathers might be better able than others to cope with environmental conditions as they change from generation to generation. But a female that is permanently turned off is not necessarily an unwilling participant. The females of some short-lived species, cecropia among them, cannot afford to hedge their bets by mating with more than one male. The life span of a cecropia female is so very short that her overriding priority must be to finish distributing her eggs before she dies.

We can safely assume that a female insect does not ponder the various facets of her recent sexual experience and on that basis consciously resolve to reject future suitors. Insects act more instinctively, their genes controlling or, at least, guiding much of their behavior. As in any animal, whatever instinctive behavior there is can ultimately be traced back to the genes and is passed from generation to generation. Furthermore, insects generally respond to only one or a few aspects of an experience rather than to its totality. Ethologists refer to such a token aspect of an experience as a "sign stimulus." Previously we saw that male silver-washed fritillaries seem to pay attention to only three aspects of the female: her size, her color, and the frequency with which she beats her wings.

Similarly, male European robins respond mainly to the red color of an intruder's breast when defending their territories. When placed in a male's territory, a stuffed red-breasted adult and a simple artificial model, consisting of nothing but red breast feathers bundled together by a piece of wire, were attacked with almost equal vigor and frequency by the defending male. But he seldom attacked a stuffed juvenile robin, which lacks the red breast but has all of the other physical attributes of a robin. The sign stimulus "red" is sufficient to trigger aggressive behavior in this bird.

There are several known answers—and certainly many that are not yet known—to the question of just which of the many stimuli provided by the male insect during copulation is or are actually responsible for abolishing the female's sexual appetite. Females of some species require nothing more than the physical sensation caused by the sperm packet in their genitalia to trigger a period during which they will reject all males. For example, it is possible to render certain unmated female cockroaches sexually nonreceptive by placing glass beads in their genitalia to simulate the presence of a packet of sperm.

In some insects, the presence of the sperm itself in the body of the female is required to inhibit her from accepting future suitors. This is the case with pomace flies, often loosely called fruit flies. Females do not lose interest in sex unless their storage pouches actually contain sperm. Virgin females that mated with intact males were turned off, but virgins that mated with castrated males were not. The castrated males were willing and able to perform, but their ejaculate contained no sperm. A similar response was discovered in cecropia females. Females that mated with castrated males continued to behave like virgins, unlike females that mated with intact males. The castrated males passed sperm packets into the sperm pouches of the females with which they mated, but they were shooting blanks; their packets contained no sperm at all.

Pheromones that emanate from males during copulation may also inhibit a female's sexual responsiveness. George B. Craig, Jr., of the University of Notre Dame, did an elegant series of experiments that show that the females of several species of mosquitoes, including the notorious yellow fever mosquito, are sexually turned off by a pheromone that the male passes into their genitalia along with his semen. Craig's experiments leave no doubt that this pheromone is produced by glands associated with the genitalia of the male. His first experiment was to dissect these glands from the bodies of males and implant them in the bodies of virgin females. (Such microscopic surgical tours de force are often performed by entomologists.) Virgin females implanted with a piece of nonglandular tissue from a male continued to act like virgins and willingly mated after the operation. But virgin females implanted with the appropriate glands from a male were matronly in their behavior and refused subsequent suitors. Accordingly, Craig named this pheromone "matrone." He then went on to show that injecting matrone-containing extracts of these glands into the blood of females had the same effect as implanting whole glands.

Not all male insects are able to alter a female's behavior so that she does not want to copulate again. In some species, females are willing to copulate many times, but males guard them after copulation or use other means to prevent later-arriving males from inseminating them. A male may even interfere with a preceding male's sperm. But these methods are not infallible. The males of many species have evolved ways of circumventing the methods that they themselves use to prevent their mates from being fertilized by other males. Thus they can cuckold other males but are, consequently, subject to being cuckolded themselves.

As already mentioned and as discussed by Kenneth Mellanby, bed bugs and many of their relatives practice traumatic insemination. The penis, enclosed in a rapierlike organ, cuts its way through the abdominal wall of the female. In some species, the sperm are injected directly into the blood-filled abdominal cavity of the female and then swim directly to the ovaries and fertilize the eggs before they are laid. This traumatic means of insemination is a shortcut to the ovaries that bypasses the usual route in other insects, in which sperm are ejaculated into the vagina, pass into a storage pouch, and fertilize the eggs as they move past the duct of the storage pouch as they are being laid. This bizarre method of insemination probably evolved as male bed bugs competed with each other to place their sperm closer and closer to the mother lode of eggs, the ovaries. Some male insects evolved long penises with which they enter the vagina but bypass the female's storage pouch and deposit their sperm farther upstream close to the ovaries. A few males, notably among the bed bugs, evolved traumatic insemination instead, and eventually this strange procedure became the norm among these insects. In some species,

including the bed bug that attacks humans, the females have even evolved special, internal abdominal receptacles, quite separate from the original genitalia, to receive the sperm.

The males of some other insect species plug the female's reproductive organs so that she cannot accept the sperm of another male even if she is willing to copulate again—much as medieval knights put iron chastity belts on their wives before they went off to the Crusades. Male migratory locusts pass into their mates a packet of sperm that doubles as a mating plug. The long, thin gelatinous capsule that encloses the sperm, like a thin strand of vermicelli about three-quarters of an inch long, is a perfect fit in the long and narrow sperm pouch of the female. For the next three days, any male that tries to copulate with a plugged female will be unsuccessful in trying to inseminate her, because he will be unable to insert his sperm packet into her sperm pouch. By the fourth day, fluids in the female's sperm pouch will have dissolved the plug, but by then there is a good chance that the first male's sperm will have fertilized all or most of her eggs. Immediately after copulation, certain male butterflies coat the genitalia of the female with a viscous, white fluid that hardens to seal her genitalia and makes it impossible for another male to copulate with her. But sometimes a second or even a third male may manage to inseminate a female. Mating pairs of these butterflies are often crowded and jostled by intruding males that try to horn in on the action. Occasionally, one or more of these males manages to mate with the female after the first male has left but before the seal has hardened. The males of some insectivorous midges have evolved a suicidal method of preventing other males from inseminating their mates. The female devours the male during copulation, but she leaves his genitalia interlocked with hers as a mating plug. His sacrifice has probably cost him relatively little in terms of fitness, measured as the number of offspring that he leaves behind. These midges mate in swarms in which males greatly outnumber females. Thus there is only a slight chance that a male could manage to inseminate a second female. He will probably gain more fitness by sacrificing himself to nourish the mother of his future progeny than by trying to mate with another female. Not only is the female midge nourished by the male's body, but she may further benefit because the male's posthumous guarding reduces sexual harassment from other males, thus giving her more time to get on with the all-important task of laying her eggs.

The males of other insects try to keep their mates faithful by daubing them with an antiaphrodisiac pheromone, a scent that mixes with the female's own scent and makes her less attractive to other males. Female mealworm beetles produce a sex pheromone to which the males are highly sensitive and to which they orient from a distance. After mating, females continue to release the sex-attractant pheromone but are much less attractive to males than they were before mating. Experiments, nicely executed by George M. Happ of New York University, prove that the first male to mate with a female applies an antiaphrodisiac to her body. In one of the experiments, an extract of a female's body, presumably containing her sex-attractant pheromone, was shown to be highly attractive to males. But the extract of the female became much less attractive after being mixed with an extract of a male's body, which apparently contained the antiaphrodisiac pheromone. Antiaphrodisiac scents are also produced by other male insects, including a tropical butterfly in which the females store the male's antiaphrodisiac for a long time, presumably in a special storage organ.

In many insect species, males that have failed to find or win over a female of their own may try to displace a successful suitor even as he courts or copulates with his partner. These interlopers tend to be very aggressive and often use brute force to separate other males from their mates. Although different male insects have evolved different strategies to cope with interloping males, there are, broadly speaking, only two options: a male may retreat to a hiding place with his mate or, as is the case in many species, he may guard his mate and try to fight off intruding males. These two options are not mutually exclusive.

Previously we saw that when males of a bumble bee-mimicking flower fly spot a female sitting on a blossom, they immediately pounce on her and carry her off to copulate in a hiding place in the surrounding vegetation. After a successful courtship between queen butterflies, the male flies off into the foliage with his passive mate dangling from his genitalia. Similarly, many other insects seek private places in which to copulate. This is almost always the case with insects that locate each other in swarms, including many gnats, midges, mosquitoes, and a scattering of others.

In 1906, Frederick Knab published a vivid description of mosquitoes swarming in Urbana, Illinois, and for the first time reported that swarming brings the sexes together. He found that a large swarm of thousands of mosquitoes formed over his head and shoulders when he walked into a recently cut field of corn adjacent to a polluted stream squirming with mosquito larvae. He then noticed that another swarm was hovering over a nearby corn-stook (a teepee-like shock of cut corn stalks). Knab went on to say: "A round of the field showed that each corn-stook had its swarm of mosquitoes, and furthermore, single stalks that remained standing had small swarms dancing over them—sometimes of only six or eight individuals." In every case the swarms consisted mainly of males, which all faced into the breeze as they hovered and danced over their chosen landmark. When he swept his net through a swarm of these mosquitoes, he caught 897 males but only 4 females. He notes that another entomologist, who swept his net through a swarm of midges, caught 4,300 males and only 22 females. Knab went on to observe that, "always the mosquitoes gathered over some prominent object such as a tree or a projecting branch, a bush, a corn-stook or a person. In this last case the swarm would move with the person and the only way to get rid of it was by passing under some taller object where the swarm would then remain." Knab reported that swarms of gnats and mosquitoes were known as far back as 1634, although their role in bringing the sexes together had not yet been discovered. He also related that swarms of midges or mosquitoes may be immense; a German entomologist reported that in 1807 the fire department was summoned when a huge swarm of gnats that formed over the steeple of St. Mary's church in Neubrandenburg was mistaken for a cloud of smoke.

Swarms are really airborne leks. As the males hover over their landmark, they wait for the unmated females that will enter the swarm in search of a mate. The males recognize the females when they hear their distinctive flight tone and immediately pounce on and couple with them. Just as quickly, the pair leave the swarm. Knab saw a female mosquito "dash into the midst of the swarm and emerge on the other side united with a male." The males apparently return to the swarm after copulating in the slim hope of mating with another female, while the females leave to get on with the business of laying their eggs on the surface of bodies of stagnant water. This, of course, explains why males always outnumber females in swarms. In some species, the pair finish copulating on the wing and soon separate. In other species, many of the dance flies, for example, the pair retreat to a hidden spot in the nearby vegetation and may remain coupled somewhat longer. Leaving the swarm during copulation probably benefits both sexes. The males gain because they avoid being displaced by another male, and the females may gain because they avoid the possible hazard and confusion of finding themselves in the center of a cluster of striving, sex-starved males.

Male insects of many species battle each other for possession of a female, or they may forcibly take over a female even as she is copulating with another male. Aggressive behavior of this sort occurs in several insect groups, but seems to be especially well developed in certain beetles. In *The Malay Archipelago*, published in 1869, Alfred Russel Wallace (co-formulator with Darwin of the theory of evolution) described a contest between two snout beetles, so called because of the long snout that projects forward from the head. Wallace wrote of his experience in the Aru Islands south of New Guinea, "two [males] were fighting for a female who stood close by . . . They pushed at each other with their rostra [snouts], and clawed and thumped, apparently in the greatest rage, although their coats of mail must have saved both from injury. The small one, however, soon ran away, acknowledging himself

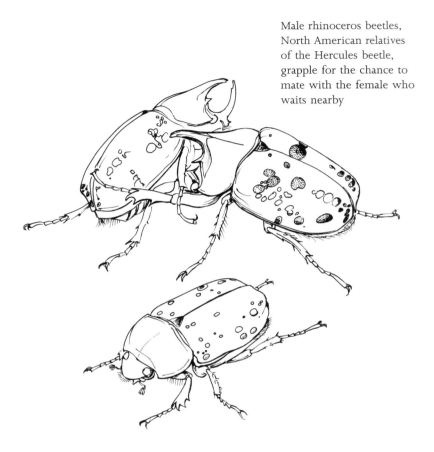

Male rhinoceros beetles, North American relatives of the Hercules beetle, grapple for the chance to mate with the female who waits nearby

vanquished." Well over one hundred years later, another naturalist described similar battles between Central American snout beetles and also reported that mateless males often displace mating males by shoving the snout between a mating pair to pry the other male off the female's back. In certain species of long-horned beetles, named for their exceptionally long antennae, lovelorn males will grab a copulating male by an antenna to pull him off of his mate's back. The victim often frustrates the attacker by holding on to the female as tightly as he can.

Some male beetles have specialized weapons that they use against other males in contests over females. These weapons may be long, oddly shaped mandibles that are obviously useless for feeding, rhinoceros-like horns, or even structures that somewhat resemble the antlers of deer. But insects, as William Eberhard has pointed out, use their horns differently than mammals do. Many horned mammals try to gore each other, although they usually do not succeed. Deer and elk lock antlers in wrestling matches that generally do not result in injury, the weaker male abandoning the field after the other has demonstrated his superior strength. Horned beetles are well armored and seldom wound each other. Unlike horned or antlered mammals, they often use their horns to pick up and carry away rivals, sometimes even to bodily lift copulating males from their mates.

Take, for example, the Hercules beetle, a fruit-eating South American scarab. The large and rotund males are giants among insects. They may be over five and one-half inches long, with their huge, forward-projecting horns constituting well over half of this length. Small males, known as minors, are only about three inches long and have horns that are roughly equal in length to the rest of the body. One of the horns, fixed in position, grows up from the thorax and arcs forward over the head. The lower horn projects forward from the top of the head. As the head moves down and up, its horn moves away from or back toward the stationary thoracic horn. Thus the two horns separate or come together like forceps. Males fight over the smaller, hornless females, which are generally less than two and one-half inches long. Like many other horned beetles, Hercules males do not try to gore each other. Instead, one tries to catch the other male between his horns, lift him up bodily, carry him well away from the female and then hurl him to the ground. Then the winner rushes back to mate with the female.

Minor males cannot possibly prevail in these contests. Why then do they exist at all? We can extrapolate a probable answer from the behavior of the South American snout beetle that was mentioned above. In this species, there are also very small minor males that never engage in combat with major males. But these little males, like silent, satellite male crickets, do sometimes manage to

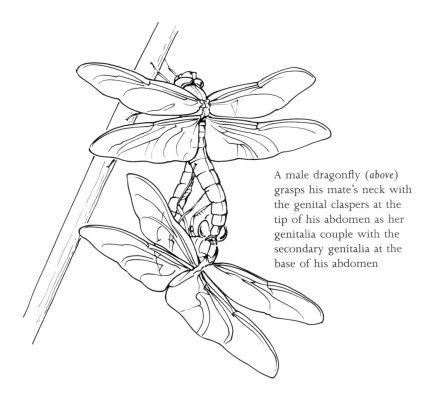

A male dragonfly (*above*) grasps his mate's neck with the genital claspers at the tip of his abdomen as her genitalia couple with the secondary genitalia at the base of his abdomen

sneak in and copulate with a female while two major males are busy battling each other for her favors.

Among those males that can interfere with the sperm deposited in a female by a preceding male are certain species of damselflies and dragonflies. Some male dragonflies have large, inflatable, membranous lobes at the tip of the secondary intromittent organ. Jonathan Waage of Brown University in Providence suggests that, during copulation, these lobes become inflated with blood and are forcefully pushed into the sperm pouch of the female. Thus the inflatable lobe packs the sperm of any preceding male deep into the back of the sperm pouch. They are then covered by the sperm of the second male, which guarantees the occurrence of sperm precedence, so that the female will use his sperm to fertilize her eggs. Many damselflies, close relatives of the dragonflies, have even evolved the ability to remove another male's sperm from the body of the female. In these damselflies, the tip of the secondary intromittent organ has intricate structures that are used to scoop out and discard a preceding male's sperm from the sperm pouch of the female. It is possible, of course, that a second male will, in his turn, be cuckolded by a third male. Some dragonfly and damselfly males have, consequently, evolved the behavior of guarding their mates for some time after copulating with them.

A copulating pair of dragonflies or damselflies seem to be locked in a wrestling hold. Their bodies form a circle as each grasps the other with the tip of its long, slender abdomen. The male's claspers, on either side of the true genital opening at the tip of his abdomen, grasp his mate by the back of her neck, an insectan approximation of a full nelson. At the same time, the female's abdomen loops forward so that the genital opening at its tip joins with her mate's secondary genitalia, located at the base of his abdomen, far forward of his true genital opening. Before grasping his mate's neck, the male prepared for copulation by looping his abdomen forward to transfer semen from his true genital opening to a receptacle associated with his secondary genitalia. When his secondary intromittent organ enters the female's vagina, it will first pack down or remove the sperm of any male that may have preceded him. Only then does he place his own sperm in his mate's genital opening. After copulation, the male releases the female's genitalia but maintains his hold on the back of her neck.

The male's secondary genitalia have freed the genital claspers at the end of his abdomen so that he can use them to hold onto his mate and guard her against rival males who would remove his sperm and replace it with their own. Males may hold onto a female for the rest of the day, but they always release her as nightfall approaches. A male may even guide his mate to the best sites for laying eggs. As you have seen, female damselflies lay their eggs under water on aquatic plants that protrude above the surface. In some species, the male holds onto the female's neck as she dips her abdomen below the water surface to lay her eggs. In

others, he releases her so that she can crawl beneath the surface but will renew his hold on her when she reemerges.

Female insects may cooperate with the males that guard them. The males of certain species of crickets normally guard their mates for over an hour after copulation. A female that is left unguarded will, after about thirty minutes, reach back with her mouthparts and withdraw the neglectful male's sperm packet from her genitalia and eat it. At that time, many of the sperm have not yet left the male's sperm packet to migrate to the female's sperm pouch. But if a male guards her for at least eighty minutes, she does not eat his sperm packet until it is empty and all of the sperm have migrated to her sperm pouch. The female thus assures that her eggs are fertilized by the sperm of a competent guarder whose sons will be competent guarders who will pass her genes on to future generations.

Some male insects may guard their mates for prolonged periods of time. But this occurs only in some species in which locating a second mate is so highly improbable that a male is more likely to profit by guarding his investment of sperm in the first female with which he copulates than by leaving her to search for another receptive female. In some species, the males stay near their mates and fight off any males that try to supersede them. But the males of other species accomplish the same end by remaining coupled with their mates for long periods, thus acting as living chastity belts.

We already know that a pair of cecropia moths remain in copula for fifteen hours, but a walkingstick holds the insectan record for prolonged copulation. The small male—as little as one-quarter the length of the female—may perch on the back of his gigantic mate's abdomen and remain in copula with her for as long as seventy-nine days. But the faithful male angler fish, the very antithesis of a male chauvinist, holds the animal world's record for prolonged devotion to his mate. It is extremely difficult to find a mate in the perpetually dark, abyssal depths of the oceans where these angler fish live. The male, therefore, does well to stay with his mate once he has found her. The tiny male, only one-thousandth the weight of a female, sinks his jaws into her skin. As time passes, his skin grows together with hers, and their blood vessels unite. His mouth degenerates and he becomes a parasite on the body of his mate, nourished by her blood and ready to release his sperm when she is ready to reproduce. He remains with his spouse for the rest of his life. When she dies, he dies with her.

After reading the first part of this chapter you may wonder whether or not the study of such seeming arcana as the sex lives of insects has any practical value. (Entomology is, after all, among other things, an applied science; just as some pharmacologists and physicians seek new drugs to kill the protozoa that cause malaria, some entomologists seek new insecticides or other more ecologically sound ways to control the mosquitoes that transmit these protozoa from sick to healthy people.) Please do not misconstrue my words. Practical application is not the only or even the major justification for pursuing entomological or any other sort of scientific research. Most good scientists do science because they love it, and many of them simply want to understand nature for its own sake. Paradoxically, it is "pure science" of this sort, exploring uncharted territory, that is most likely to open up entirely new avenues of research that ultimately make possible previously undreamed of practical applications. No matter how you feel about nuclear bombs and whether or not you approve of nuclear energy, you must concede that the intellectual efforts of Niels Bohr, Albert Einstein and other physicists to satisfy their curiosity about the structure of the atom and the fundamental nature of matter have had momentous practical consequences. Similarly, the work of the Nobel laureates James D. Watson and Francis H. C. Crick in unraveling the structure of the DNA molecule, the stuff of the genes, launched the study of molecular biology. They thus revolutionized our understanding of life and made possible such practical applications as gene therapy and the mass production of human insulin and growth hormone by inserting human genes into bacteria.

The answer to the question of whether or not understanding the sex lives of insects has had any practical value is a resounding yes. Such knowledge has been useful in many ways. As you already

know, sex-attractant pheromones can be used to monitor or even control noxious insects. But more germane to the subject of this chapter, the knowledge that male insects can turn off the sexual behavior of females made possible the most spectacular solution to an insect pest problem that has ever been achieved.

The object of that spectacularly successful control procedure, the screwworm fly, named for the screwlike ridges on the body of the maggot, is one of the most horrifying of all of the pest insects. Its scientific name, *Cochliomyia hominovorax*, means literally "the snail-like fly that devours people." Screwworms do sometimes attack people, but they more often attack cattle and wild animals such as deer. The females lay masses of from 200 to 400 eggs near wounds on these animals, perhaps a nick from barbed wire or the bloody navel of a newborn calf. The maggots that hatch a day later enlarge the wound as they devour the healthy flesh at its edge. As the wound enlarges, it attracts more and more egg-laying females and it grows ever larger. Without intervention, the result is often death. Screwworms once attacked tens of thousands of cattle in the United States each year. The only cure was to search the range for each infested cow and to smear its wounds with an insecticidal ointment, a prohibitively expensive procedure. The occasional human cases were gruesome. A text on medical entomology shows a photograph of a man whose face was almost entirely eaten away by screwworm maggots as he lay unconscious in a field for several days.

In 1992, Edward Knipling and Raymond Bushland, both now retired from the U.S Department of Agriculture, shared the $200,000 World Food Prize for their development of the sterile insect technique for controlling pests. This environmentally friendly method, which does not involve the use of any insecticides, made it possible to *permanently eradicate* the screwworm fly from North America south through Mexico and Guatemala. It will soon be eradicated all the way to Panama and may eventually be eradicated from South America. Such an accomplishment was once considered impossible, more in the realm of science fiction than science. It has been tried with insecticides, but that never worked. No insect has ever been eradicated from more than a small area by the use of insecticides.

When Knipling and Bushland set out to control this ghastly pest, they thought that they were more likely to succeed if they went back to fundamentals than if they tried to rehash known control techniques such as the use of insecticides. One of their early fundamental research projects was a study of the sex life of the screwworm fly. This could sound a bit ridiculous to the uninitiated, couldn't it? But fortunately, Senator William Proxmire did not jeopardize government support of their study by presenting Knipling and Bushland with one of his infamous, self-proclaimed Golden Fleece Awards, meant to ridicule scientific research that Proxmire, trained in business administration, did not understand and considered to be frivolous and of no practical consequence.

Knipling and Bushland soon discovered that screwworm males mate many times, but that the females are sexually turned off by mating with only one male. They then asked some questions that no one had ever asked before: Is it possible to sterilize screwworm males, and will they afterward retain their interest in mating? Would mating with a sterile male turn off a female's interest in sex and thus condemn her to laying unfertilized eggs? Would it be possible to reduce or even eliminate screwworm populations by releasing into their environment sterile males that would compete for females with the wild, fertile males? The answer to all of these questions turned out to be yes.

Males, quickly and easily sterilized with radioactive cobalt (another application of nuclear physics), remained sexually active, and females that mated with a sterile male never mated again and laid only infertile eggs. The last question was brilliantly answered on the island of Curaçao in 1953. After setting up "screwworm factories" to produce sterile males, Knipling and Bushland released several succeeding waves of sterile males on the island. The results were spectacular. So many of the wild females mated with sterile males and laid infertile eggs that screwworms were eradicated from Curaçao in only a few months. They have not reap-

peared there in forty years. The screwworm was next eradicated from peninsular Florida. During the winter of 1958–59 hundreds of millions of sterile males were released from aircraft on 85,000 square miles of southern Florida. (No one was ever bothered by them because they were released at the rate of only 200 per square mile per week.) The last screwworm ever to be seen in Florida was found on February 19, 1959. This control program cost $10 million, but at that time the yearly loss to screwworms in Florida alone was $20 million, which is about 140 million in inflated 1992 dollars. Every year, winter weather had wiped out all of the screwworms in the southeast except for the population in southern Florida, but every summer they had reinvaded the rest of the southeast and often appeared as far north as Illinois. Thus eradicating these pests from Florida freed the entire southeast from them. The subsequent eradication of screwworms from the southwest freed the whole country of these damaging flies and has to date saved our economy billions of dollars. Political problems have slowed the progress of screwworm eradication, but when these flies have been eliminated all the way south to the 60-mile-wide Isthmus of Panama, a project that is under way now and should be completed by the time this book is in print, the South American population will be prevented from invading Central and North America by the creation in Panama of a narrow, and therefore cheap to maintain, barrier zone, a "no man's land" on which sterile male screwworms will be frequently released. Some day the screwworm will be eliminated from all of South America as well as North America. Then no barrier zone will be necessary. The story of the eradication of the screwworm is summarized by R. L. and R. A. Metcalf in their authoritative book, Introduction to Insect Pest Management.

Caring for Offspring

Although we do not tend to think of insects as caregivers, it seems inevitable that they, like so many other animals, would evolve the capacity to bestow parental care and thus improve their progeny's chances for survival. After all, the fitness of any parent depends not upon the total number of offspring that it produces, but, rather, upon how many of them survive to reproduce themselves and thus project copies of their parents' genes into the future. These gene copies can program the development of grandchildren and great-grandchildren that may survive the gauntlet of natural selection to pass this genetic information on to even more distant descendants.

The season is now early summer. The solstice fell on June 21, less than a week ago. In the Northern Hemisphere, the summer solstice is the longest day of the year and the sun is at its highest. Paradoxically, the hottest part of summer will come during the shorter days of July and August, when the earth radiates the heat that it has been absorbing since spring.

Remarkable changes in the flora and fauna have taken place since early last April, when the cecropia pupa in the cocoon beneath the birch began its metamorphosis. From then until May, most of the wildflowers in blossom were woodland species taking advantage of the sun before the trees formed the leafy canopy that now shades the forest floor. At that time, columbine, Mayapple, Dutchman's breeches, spring beauty, and other heralds of spring supplied nectar and pollen for flies and solitary bees, many of which occur only when these and other woodland wildflowers are in bloom.

By early summer, wildflowers have become scarce in the woodland shade, but they are much more abundant in open places. In a clearing in the woods, a trumpet creeper twines on the bole of a dead oak. As they hover in front of its deep, orange blossoms, ruby-throated hummingbirds probe for nectar with their long tongues. That evening, a hawk moth slides its equally long tongue into the nectar-rich blossoms of a nearby honeysuckle. Along roadsides, the lacy yellow inflorescences of wild parsnip swarm with nectar-seeking flies, and the lovely, pale blue flowers of wild chicory—open only in the morning—are visited by honey bees. In meadows, worker bumble bees collect nectar from red clover, and a variety of insects visit the blossoms of black-eyed susan, ox-eye daisy, and common milkweed. Among these visitors are such butterfly sprites as the American copper, the silvery checkerspot, and the pearl crescent. Among the most beautiful and noticeable butterflies at this time of year are the first few, newly emerged monarchs. Fresh and bright, they are the offspring of the worn migrants (or of their children) that returned from Mexico earlier in the year.

The cecropia moths that began to metamorphose in early April and emerged in May disappeared long ago. Even the bodies of the few that survived to die of old age have been eaten by ants or other scavengers. The surviving offspring of these early moths are now partly grown caterpillars that will spin their cocoons and metamorphose to the pupa in less than two months. The late-emerging cecropia moths, which did not start to metamorphose until June, began to appear about two weeks ago, and they will continue to emerge from their cocoons for yet another week. As did the cecropias of the early group, the males search for virgin females, and the mated females are busy laying their eggs. The caterpillars that hatch from them will not spin cocoons until September.

Like cecropia caterpillars, other young animals are doing their best to survive and grow to maturity—some on their own and some with help from their parents. Half-grown grasshoppers graze on broad-leaved plants and grasses. The eggs from which they came, laid in the soil during the previous autumn by their now long dead mothers, were in diapause until shortly before they hatched just a few weeks ago. Brown thrashers are about to turn loose their first brood of newly feathered young. In July they will lay a second clutch of eggs. Blue jays have fledged their one and only brood for the year. The young jays and their parents will eat almost anything, sometimes the eggs of another bird and sometimes even a large cecropia caterpillar. (One of my graduate students once saw a blue jay in flight pluck a full-grown cecropia caterpillar from a twig with its beak. Since the jay probably weighed only about three ounces and the caterpillar almost an ounce, this was equivalent to a 180-pound man snatching a 50-pound weight with his teeth as he swung past on a rope.) A white-footed mouse suckles her young inside her domed nest, which may be in a burrow in the soil or even in an abandoned thrasher's nest in a tangle of briars. (Next summer, her abandoned burrow may serve as home for a colony of bumble bees.) In summer, her diet consists mostly of insects, including any cecropia larvae that she happens to come upon. During the winter, she will subsist mainly on seeds but she will also eat some insects, including the pupae in cecropia cocoons.

Many animals, particularly the more primitive ones, offer their offspring no care at all. They provide them with the bare essentials, nothing more than what is contained in the egg. Their eggs, like those of most animals, contain the genetic blueprints which ensure that the developing embryo will become a copy of its parents, the nutrients required for the growth of the embryo, and, in some species, extra yolk to sustain the embryo for a short while after it hatches from the egg. Such animals simply release their eggs into the environment more or less at random and invest no more time or energy in them. As a general rule, they lay many small eggs.

Wood ticks and dog ticks, distant relatives of insects, spew a single mass of from 4,000 to 7,000 eggs on the ground. The newly hatched larval ticks must, all on their own, find and board a mammal on whose blood they can engorge. Parasitic round-worms of humans, common in the tropics but generally uncom-mon in North America and Europe except on pigs, may survive in the intestines of a host for many months, and an adult female may lay 200,000 eggs per day that are passed out with the feces of the host. An egg's chance for life is very slim indeed. It will live only if it is inadvertently ingested by a human or a pig, sometimes when a person eats a poorly washed raw vegetable that was fertilized with human excrement. The members of such exceedingly prolific species as oysters, ticks, and roundworms play the odds, betting that some of their many eggs—the more the better—will, despite being abandoned, grow up to carry on their genetic line.

Other species produce fewer offspring but improve the odds for their survival by giving them some measure of parental care. The magnitude of this additional investment of time and energy varies from animal group to animal group and from species to species. As a general rule, the number of eggs laid decreases as the level of care increases and as threats from the environment decrease.

Parental care is so beneficial that it comes as no surprise that many species in several unrelated groups, including the insects, have invented it quite independently of each other. Birds and mammals are famous for the care that they lavish upon their young. Human maternal care is a recurrent theme in the arts. Take, for example, the many paintings of the Madonna and Child. In one of the loveliest, Jan van Eyck's *Lucca Madonna*, which hangs in the Städelkunstinstitut in Frankfurt, we see Mary offering her breast to the infant Jesus.

The really amazing thing is, however, that simple forms of parental care occur in even such primitive animals as jellyfish, which are so low on the evolutionary scale that they do not even have respiratory, circulatory, digestive, or excretory systems. These functions are carried out in the primitive, all-purpose body cavity, whose only opening to the outside world is the "mouth." Testes and ovaries empty into this cavity. The male passes his sperm out into the sea through his mouth. A female retains her eggs in her body cavity, where they are fertilized by sperm that are carried in through her mouth with the water that comes in when she ingests food. Then she passes the fertilized eggs out through her mouth but immediately gathers them together in folds of tissue that surround the mouth on the outside. There she broods them in safety until they are ready to hatch.

Other animal groups, groups that most people do not associate with parenthood, actually do include exceptional species that care for their young. Although oysters and many other mollusks offer no parental care, freshwater clams brood their eggs in their body until they have hatched and passed through all but the last larval stage. Then they are released, and those that are lucky enough to find a fish live as parasites on its gills, fins, or skin until they become mature and drop off to burrow into the bottom sediments. As we will see in the next chapter, some species of freshwater clams do their offspring the additional service of releasing them only after the mother clam has attracted a fish to serve as their host. Octopuses, the most intelligent of the mollusks, lay their eggs in their dens and guard and clean them until they hatch. Their offspring, which do not pass through larval stages, are miniatures of their parents and are ready to fend for themselves upon hatching.

Seahorses are not the only fish that care for their young. Sharks brood their eggs in their bodies until after they hatch. *Tilapia*, now being raised on farms as a food fish, is a mouth brooder. The female spawns in a nest excavated by the male. But immediately after the male covers her eggs with sperm, she takes them up in her mouth and holds them there in safety until they hatch. The young remain near her and retreat into her mouth en masse if danger threatens.

Javanese flying frogs, so called because their webbed feet are broadly expanded for gliding, build nests that keep their tadpoles out of predator-infested waters for at least a while. As the eggs are being laid, the male and female together build an aerial, hanging nest of leaves and whipped-up foam that the female produces with the eggs. While the outside of the nest hardens, the inside liquifies to form a pool in which external fertilization occurs and the tadpoles swim as they are kept alive by the nutrients stored in their yolk sacs. The next beating rain will wash them down into the body of water over which the nests always hang and in which predators often lurk.

⟍ Many insects give their young some form of parental care, but others provide only the most perfunctory care or none at all. For example, some mayflies simply discharge their approximately 3,000 eggs into a river or lake where they sink to the bottom, and the aquatic nymphs must shift for themselves. Some species of mayflies do not even lay their eggs. The whole abdomen with the eggs that it contains simply breaks off and falls into the water. A female trigonalid wasp lays about 10,000 eggs that she sticks indiscriminately to the leaves of various plants. An egg can survive only if it is inadvertently swallowed by a caterpillar that eats the leaf to which the egg is attached. The larvae that hatch from the eggs of some trigonalids parasitize the caterpillar itself. Other species survive only if the caterpillar that eats the egg is internally parasitized by some other insect. The larvae of these trigonalids can live only in the body of the parasite of the caterpillar. The larvae of yet other trigonalid species survive only in the bodies of the larvae of caterpillar-eating wasps such as the colonial yellowjackets. In this case, the trigonalid egg will not survive unless the caterpillar that swallows it is caught by a predaceous wasp that feeds it and the egg that it contains to one of her larval sisters. The odds that a trigonalid egg will be swallowed by a caterpillar—let alone one that is already parasitized or one that will later be caught by a wasp—are so slim that it is no wonder that the females must lay so many eggs. The interested reader should consult K. W. Cooper's summary of the habits of trigonalids.

⟍ A few insects retain their eggs within the body until embryonic development has been completed, at which point they give live birth to their young. These newly born individuals must still cope with the world outside of the "womb" on their own during their nymphal or larval stages, but at least they are spared exposure to parasites, predators, and the elements during the egg stage.

Aphids of most species survive the winter as diapausing eggs—the rosy apple aphid, for example, in a crevice in the bark of an apple twig. In the spring, every last one of these eggs that hatches will produce a female that will, when she matures, reproduce without benefit of a male (a process called parthenogenesis) and give live birth only to daughters immediately after they complete embryonic development. During the spring and summer, there are several generations of these aphids, all females, all parthenogenetic, and all live bearers that give birth to young at the rate of about seven per day. The embryos in the body of a live-bearing mother already contain early stages of the embryos of the next, or third, generation. Aphids can become pregnant even before they are born, and the mother carries both her children and her grandchildren within her "womb" at the same time. Thus there are aphids within aphids within aphids. In late summer or fall, parthenogenetic females of the penultimate generation give live birth to a final generation that consists of both males and females that will mate and lay the eggs that will survive through the winter. Dispensing with "useless" males during spring and summer probably doubles the reproductive rate, helping to make up for the depredations of the many parasites and predators that plague aphids.

In a few insect species, the female retains not only the egg within her body but also the larva or nymph, much as humans and other mammals carry the fetus. In some species, the larva lives and grows within the mother's body until it reaches full size and is ready to transform to the pupal stage. Thus it is spared the dangers presented by the outside world. Among these insects is the sheep ked, a wingless fly that lives as a blood-sucking parasite in the fleece of sheep. Another such insect is the tsetse, a fly that we met briefly before. Tsetses are immensely important in Africa because they transmit both sleeping sickness, a debilitating disease of humans, and nagana, a similar disease of cattle. Tsetses, consequently, have been much more thoroughly studied than sheep keds, which have only a minor economic impact on people. P. A. Buxton relates that female tsetses produce young one at a time in a manner strikingly similar to gestation and birth in mammals. The egg hatches internally, and the larva is retained in an analogue of the mammalian uterus. Instead of the placenta that nourishes fetal mammals, a milk-secreting gland in the "uterus" feeds the larva. Tsetse milk is white like cow's milk and chemically and nutritionally similar, containing fats, proteins, vitamins, and minerals. After about ten days, the larva is fully grown, and the mother retreats to a shaded place where she gives birth to an offspring that is almost as large as she is. David L. Denlinger of Ohio State University and Jan Zdárek of the Czechoslovak Academy of Sciences, working together at the International Centre of Insect Physiology and Ecology in Nairobi, Kenya, measured and described in detail the contractions that female tsetses undergo when they expel a larva from the "uterus." Immediately after it is born, the larva digs into the soil and transforms to the pupal stage. About three weeks later, the adult emerges from the pupa and flies off to feed and locate a mate. As is to be expected, tsetses produce far fewer young than do insects that do not give comparable parental care. A female kept alive in a laboratory for an unusually long time gave birth to a record total of twenty larvae. In nature, tsetses live for a much shorter time and, as reported to me by David Denlinger, probably produce an average of only two to three larvae during a lifetime.

Most insects lay their eggs long before embryonic development has been completed, but many of these species make some provision to safeguard them. Some lay eggs whose appearance tends to protect them from attacks by predators or parasites. Many lay inconspicuous, camouflaged eggs that blend in with the background and thus tend to escape notice. But a giant silk moth in Mexico lays conspicuous eggs, each with a round black marking that looks like the hole that an emerging parasite makes in an eggshell. Foraging birds may be tricked into passing up these eggs, because they have learned that eggs with escape holes have already been emptied by parasites.

The bodies of certain caterpillars are covered with poisonous, needle-sharp hairs that, like nettles, cause an irritating and sometimes agonizing pain if they brush against the skin. Although adult moths do not grow poisonous hairs, females of some species salvage them from their own cast larval skins before they emerge from the cocoon. The poisonous hairs are entangled in a tuft of barbed, nonpoisonous hairs at the tip of the female's abdomen. When she lays a mass of eggs, she covers it with poisonous hairs to give it the same protection from predators that the hairs gave her when she was still a caterpillar.

In one of his 1790 volumes on the moths and butterflies of Europe, the German entomologist Moritz Balthasar Borkhausen told how the brittle, poisonous hairs of processionary caterpillars were from ancient times ground up and used to commit murders by poisoning. He describes the effects on the victim in gruesome detail. The drinking of a poisoned draft is followed by sharp pains in the lips and gums. Then come agonizing pains in the tongue and intestines, which are soon followed by vomiting, violent convulsions, and death.

Many other insects give their eggs rudimentary care by hiding them in one way or another. Some of the walkingsticks, or stick insects, well known for their camouflaging resemblance to twigs, bury their eggs in sand or soil or, as may the two-striped walkingstick of the southern United States, in the punky wood in a rotting tree stump. Burying the eggs protects them from tiny, parasitic cuckoo wasps that run about on the forest floor laying their eggs in the much larger eggs of walkingsticks. The cuckoo wasp larva lives in and eats the walkingstick egg, ultimately destroying all but the shell.

Other stick insects, including a species that is common in the northern United States, simply drop their eggs to the forest floor as they move about feeding on the leaves of trees. These eggs would seem to be easy victims for cuckoo wasps, but many of them are rescued by ants that are attracted to them and carry them off to their underground nests. Early in the 1990s S. G. Compton and A. B. Ward of Rhodes University in South Africa and L. Hughes and Mark Westoby of Macquarie University in Australia, working with other species of walkingsticks, independently discovered that ants do not eat walkingstick eggs, but that they do eat a large edible appendage of the egg called the capitulum. Eggs with the capitulum removed by ants had not been injured and hatched normally. The ants did not molest the hatchlings, which presumably made their way up into the trees. Many plants have evolved similar methods of providing for their seeds. Their seeds bear edible appendages, called elaiosomes, rich in fats and proteins, that are also eaten by ants. The ants carry these seeds down into their nests and eat the elaiosomes but do not injure the seeds themselves. Thus they protect and disperse these seeds.

Like the cuckoo wasp that parasitizes walkingstick eggs, most of the thousands of other insects that are parasites of insects place their eggs in or on the body of the host. Parasitic flies, such as the tachinid that spends its larval existence within a cecropia caterpillar, often glue their eggs to the skin of the host. Females of the family known as the thick-headed flies have an egg-laying organ (called an ovipositor by entomologists) that is shaped somewhat like an old-fashioned, prying type of can opener. It is used to place an egg within the host of their larvae, usually an adult bumble bee or nonparasitic wasp. This is unusual for two reasons. Most flies do not have ovipositors that can pierce, and most parasitic insects live in larvae or nymphs rather than in adult insects.

Parasitic wasps have sharp, piercing ovipositors that they use to inject their eggs into the body of the host. (In bees and many nonparasitic wasps, the ovipositor has been converted into a stinger for injecting venom and is no longer used for laying eggs.) These parasites tend to be quite specific about their choice of hosts. Some tiny species lay their eggs only in the eggs of other insects, often those of moths or butterflies. Other almost equally small wasps lay their eggs in the bodies of aphids. Many species prefer caterpillars, the grubs of beetles, or the immature forms of insects of other kinds. The larvae of *Megarhyssa*, a wasp with no common name, parasitize larvae of horntails, sawflies that burrow in the wood of dead trees at a depth of half an inch or more beneath the surface. The horntail larvae seem inaccessible, but *Megarhyssa* has a formidable ovipositor that may be as much as three inches long, about twice the length of her body. Although the ovipositor is as thin as a horse hair, it can drill through solid wood to place an egg in the tunnel of a horntail. Most other parasitic wasps have shorter ovipositors. They parasitize insects that are exposed or at least not so deeply hidden.

Some parasitic insects shift the burden of finding a host to their newly hatched larvae. The larvae of certain blister beetles are nest parasites of solitary bees. They live in the nests of these nonsocial bees and eat their eggs, larvae, stored pollen, and honey. The female beetles of one group of these nest parasites lay their eggs in holes that they dig in the soil. The newly hatched larvae, tiny but extremely active, scurry about and climb up onto plants, where they settle down on blossoms. If a solitary bee visits the blossom, the parasitic larva grasps the bee's hairy body and climbs on board for a free ride to the bee's nest. The females of another group of blister beetles spare their larvae part of the onerous task of finding a host by laying their eggs directly on the blossoms, where their larvae will await the arrival of a solitary bee. In either case, the probability that a larva will end up in a bee's nest is slim, and needless to say, blister beetle nest parasites lay many hundreds or even thousands of eggs, far more than do parasitic insects that lay their eggs directly on or in the host.

When they are maggots, some flies live as parasites in the bodies of mammals: horse bots by the hundreds in the intestines of horses; sheep bots by the score in the nasal and frontal sinuses of sheep; and ox warbles singly in large boils on the backs of cattle. The adult females of these species lay their eggs directly on the body of the mammal that the larva will parasitize.

But the human bot fly of Central and South America, also known as the tórsalo, uses an indirect means of getting its eggs to the host. The female tórsalo does not approach the host. Instead, she subdues a mosquito or some other blood-sucking fly and glues a small mass of eggs to the underside of its body. When the blood-sucking insect later lights on a human—or sometimes

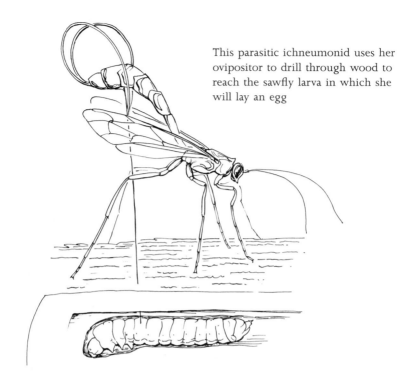

This parasitic ichneumonid uses her ovipositor to drill through wood to reach the sawfly larva in which she will lay an egg

another kind of mammal—the bot fly maggot immediately pops out of the egg and burrows into the skin, where it begins to consume the host's tissues.

In 1929, Lawrence H. Dunn, a medical entomologist at the Gorgas Memorial Laboratory in Panama, placed two just-hatched tórsalo maggots on his own skin in order to observe their behavior and their effect on him. Dunn wrote:

October 9—At the end of 42 minutes after this larva was placed on my arm it was completely out of sight in the skin . . .

October 24—The lesions [caused by the larvae] have increased in size and have . . . the appearance of small boils which have . . . erupted slightly.

November 3—At about seven o'clock this evening when the bandages were removed for a while this larva was found to be pushing out a cast skin . . . it probably needed more room in the hole [in my flesh] for its future growth . . .

November 22—The pain from the lesion becomes so intense at times that it is almost impossible to walk until the paroxysm is over.

November 29—While I was taking a shower this morning, larva no. 1 protruded for nearly half an inch for a few moments, when it again withdrew. Ten minutes later it again began to come out . . . The emergence covered a period of about one-half hour. The exact infestation period was 50 days and 15 1/2 hours.

If Dunn had been out of doors, larva number 1 would have fallen to the ground and burrowed into the soil to form a chamber in which it would have molted to the pupal stage and ultimately metamorphosed to the adult stage.

Many plant-feeding insects give their young an extra boost by placing their eggs on the species of plants that their newly hatched offspring will eat, often only a few closely related species in the same family of plants. As a general rule, these accommodating mothers lay fewer eggs than the thousands that are laid by many other species that simply release their eggs more or less at random. For example, a cecropia female lays only about 350 eggs during her lifetime, hawk moths may lay about 400; and fruit flies such as the apple maggot may lay from 300 to 400. But more about botanically specific feeding in a later chapter.

Cecropia females lay their eggs on many unrelated species of plants that belong to many families. But they are choosy in another way. With only one known exception, the garden peony, cecropia caterpillars under natural conditions eat only the leaves of broad-leaved, woody plants: trees and shrubs. They do not feed on conifers or on nonwoody, herbaceous plants such as the dandelion or daisy. Cecropia eggs and caterpillars can be found on diverse plants from many families: wild black cherry, sycamore, and various species of birch, dogwood, maple, and willow. No

one has discovered how cecropia females tell woody from non-woody plants. It may be advantageous for cecropias to limit their menu to woody plants, but no one knows for certain just what that advantage might be. It could be that most woody plants, unlike most herbaceous plants, are large enough to support a cecropia caterpillar throughout its life. A small herbaceous plant might be all eaten up before even one huge cecropia caterpillar could complete its growth on the small amount of food that the plant offers.

Cecropia can be called a "fugitive species" because its population is constantly shifting to recently disturbed areas, abandoned crop fields, or places that were denuded by a landslide or forest fire. Such places are attractive to ovipositing cecropia females when they are covered by shrubs and young trees. Later in the succession, when the area is clothed by more mature forest, it becomes much less attractive. Since disturbed areas tend to be widely separated, cecropia females presumably fly long distances to find them.

No one has followed a female cecropia through the nights of her life to find out how widely she ranges as she disperses her eggs. That would probably be impossible. She flies faster than most people can run. With flashlights, Jim Sternburg and I watched newly emerged females as they began to lay eggs just after dusk on the first night of their adult lives. Most of them laid their first clutch on a stem or a leaf of the shrub on which they had fed as caterpillars. We followed some of them as they flew to nearby shrubs of the same species to lay a few more clutches. But after a short while, all of the moths that we watched flew out of sight. There is no way to know how far they flew on that night or on succeeding nights.

Some insects give their offspring considerably more overt care than do those that I have discussed thus far. Certain crickets, bugs, and beetles guard their eggs and then continue to guard their nymphs or larvae after they have hatched. In a 1915 issue of the *Bulletin* of the U.S. Department of Agriculture, David E. Fink related how lace bug females guard their young and described their solicitous, henlike behavior when the sap-feeding nymphs have sucked one leaf dry and must move on to a fresh leaf: "When migrating from one leaf to another the female adult usually directs the way, and with her long antennae keeps the nymphs together or rebukes any straggler or deserter. It is an interesting sight to observe the migration of a colony of more than a hundred nymphs, with the female adult hurrying from one end of the flock to the other, keeping them together, and at the same time urging them in the right direction during the migration." Fink goes on to report how a female protected her brood from an approaching ladybird beetle that probably intended to eat them: "with out-stretched, slightly raised wings [she] suddenly darted toward the intruder, driving it away from the leaf."

Shortly after she first mates, the female short-tailed cricket of the southeastern United States digs an underground burrow in which she lays her eggs and cares for her nymphal offspring until they are partly grown. She stocks the burrow with food, bits of vegetation from the surface, and aggressively defends it against intrusions by either males or females of her own or other species. Like cecropia females, but unlike other female crickets, she mates only once. About a month after mating, she begins to lay eggs in the brood chamber at the end of her burrow. As each egg is laid, she picks it up gently and places it on a pile with the others. After she has completed her clutch of about twenty-five eggs, the female continually picks up and mouths each egg, apparently cleaning it to prevent mold infestations. The nymphs eat the vegetable matter in the burrow as well as an extraordinary "baby food" produced by the mother. She lays miniature eggs, probably unfertilized, that serve as food for the nymphs. The nymphs seem to be passionately fond of these miniature eggs, clustering around them and vying for the opportunity to eat them.

Quite a few other insects lay their eggs in nests that they build and stock with food for their offspring. Among them are certain dung beetles, and almost all of the bees and predaceous wasps. The nests of dung beetles are always in the soil. Bees or wasps may place theirs in a natural crevice, a hollowed-out stem, a structure that they build of mud or paper, a burrow that they dig

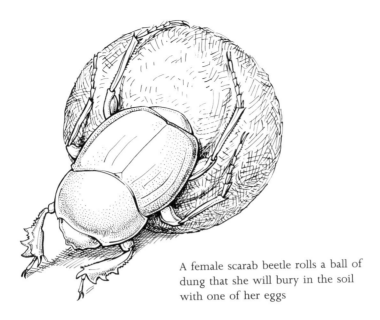

A female scarab beetle rolls a ball of dung that she will bury in the soil with one of her eggs

in the soil, or a tunnel that they bore in timber. Dung beetles stock their nests with balls of dung, bees with a mixture of pollen and honey, and predaceous wasps with spiders or various kinds of insects.

Among the dung beetles that build nests and provide food for their offspring, the best known is certainly the sacred scarab of ancient Egypt, appropriately dubbed with the scientific name *Scarabaeus sacer*. When a female is ready to reproduce, she visits a cow pat and, using her legs, forms a spherical ball of dung. The ball is larger than she is, but holding it with her hind legs and walking backward she rolls it away to a patch of soil that is suitable for digging the nest. There she digs a short tunnel with an underground brood chamber at its end. She then seals the ball and a single egg in the chamber. The larva eats the dung, metamorphoses to the pupa in the chamber, and eventually emerges from the earth as a full-grown beetle.

The scarab was sacred to the ancient Egyptians because they saw cosmic symbolism in its anatomy and behavior. To them, the beetle, with a halo of spines radiating from its head, represented the rising sun. The dung ball represented the earth. The five segments of each tarsus (foot) of the beetle's six legs add up to thirty, the number of days in a solar month. In a 1987 article in the *Revue de l'Histoire des Religions*, Yves Cambefort of the Musée National d'Histoire Naturelle in Paris argued that the significance of the sacred scarab to the ancient Egyptians might have been even greater than it has been thought to be. He suggests that the Egyptians associated the scarab with Osiris, lord of the underworld, ruler of all the dead who live there and the god of fertility who presides over the sprouting of grain and the emergence of other life from the earth. To the Egyptians, the metamorphosis of the mummy-like pupa to a new beetle may have symbolized victory over death. Cambefort suggested that the pyramids with their included burial chambers are actually idealized dung pats. This makes a sensational story but seems dubious when we consider that the ancient Egyptians doubtless knew that scarabs do not make their brood chambers in dung pats.

Although parental care is more highly developed in the social wasps and in the social bees, the solitary, nonsocial bees and the solitary, predaceous wasps also build nests and provide food for their offspring. The nests of leaf-cutter bees are well hidden, but the holes that they make by removing pieces from the leaves of various plants, especially the rose and the redbud, are obvious and familiar to many of us. These pieces, always removed at the margin of the leaf, are either oblong or circular. The circular ones are somewhat less than the size of a dime and the oblong ones are about twice as big. The bees use these leaf pieces to build thimble-shaped cells for their larvae in holes that they have found, dug in the soil, or bored in timber. Several oblong pieces are used to form the bottom and the sides of a cell. The completed cell is partially filled with a paste of pollen and honey, on which a single egg is laid. The cell is then tightly capped with circular pieces of leaf and is given no further attention. Usually, several cells are placed end to end in one burrow.

Most of the solitary wasps are quite choosy about the prey that they provide for their offspring. Different species specialize in hunting for spiders, cockroaches, aphids, bees, leafhoppers, caterpillars, or other insects. The cicada killers are so called because they provision their nests only with the nongregarious dog day cicadas that we hear droning in trees in July and August—long after the periodical cicadas have come and gone. Before hunting for her prey, the female—a giant among wasps at a body length of two inches—digs a burrow in the soil that is about half an inch in diameter and that may extend ten inches below the surface. At its end, and eventually at the ends of branches of the original burrow, she excavates the large, oval chambers that will receive her prey and house her larvae. Then she flies off to hunt for cicadas in nearby trees. (Since two-thirds of her prey are the mute females rather than the noisy males, it appears that she hunts by vision rather than by sound.) She stings her prey with a venom that keeps them in deep paralysis for some time before they die. Thus the cicadas do not decompose before the cicada killer larva can eat them. When the female has placed from one to three cicadas in a chamber, she lays an egg on one of them and then seals the chamber. Her offspring remain in their chambers until the following spring, when they will emerge as adults in late June or July.

Female cicada killers hunt in trees that may be more than a hundred yards from the nest. How do they find their way back? Departing females make circling flights over the area of the nest, possibly for the purpose of memorizing nearby landmarks. Whether or not cicada killers orient to landmarks is not known, but Niko Tinbergen, already familiar to you from his work with the grayling butterfly, demonstrated that another solitary wasp, the bee-killing *Philanthus* of Europe, does indeed use landmarks to relocate its nest. Before a *Philanthus* female flies away from her nest, she often makes an "orientation flight" similar to the one made by cicada killers. Tinbergen designed a simple but elegant experiment that proved that the females memorize landmarks near the nest. When a female was in her burrow depositing a paralyzed bee, he placed a circle of pine cones around the nest entrance and left them there while she made several trips back and forth from the nest. At first her orientation flights were longer than usual; the sudden appearance of the pine cones seemed to confuse her. When Tinbergen was convinced that the female had become accustomed to the presence of the pine cones, he moved them several feet to one side of the nest while she was off searching for prey. When she returned, she did not go to the nest entrance but landed in the middle of the displaced circle of cones and futilely searched for a nest entrance that was not there. Tinbergen repeated this experiment many times. The wasps were fooled every time.

Like most solitary bees and most of the nest-making wasps, leaf-cutter bees and cicada killers are "mass provisioners": they prepare and complete the cells in the nest one by one. They pack each cell in the nest with all of the food that the larva will need during its lifetime. Then they lay an egg in the cell, seal it shut, and pay no more attention to it. By contrast, some of the solitary wasps are "progressive provisioners," which care for several cells at one time and add food daily to each cell as the larvae grow. A cell is sealed only when the larva is fully grown and ready to molt to the pupal stage. The larvae in the cells differ from each other in size and in the quantity of food that they require. Eggs and full-grown larvae need not be fed. A newly hatched larva will need only two or three small insects, while a large larva will need many more.

How does a provisioning female manage this complex situation? In 1941, a Dutch ethologist, G. P. Baerends, considered this question in his research with a digger wasp that nests in sandy soil and feeds its larvae caterpillars. One female may care for as many as fifteen offspring at the same time, each one in its own separate nest in the soil and at a different stage of development from the others. Not only does she remember the location of each of her nests, but she can determine and remember how many caterpillars she must add to each nest each day in order to satisfy the needs of her offspring.

In the morning the female makes the rounds of her nests and inspects each one. Later in the day she returns to each nest and takes action appropriate to the condition of that nest during her morning inspection. When she returns to a nest in which she had earlier seen a full-grown larva spinning its cocoon, she comes without food and seals the nest shut. She brings two or three small caterpillars to a nest in which she had seen a small larva, and to one in which she had found a large larva she brings a much greater quantity of food. The female digger wasp is obviously capable of complex learning. On her morning visit she memorizes the condition of each nest, and on her later visit she acts accordingly. But by manipulating nests, Baerends found that the female is capable of judging the condition of a nest only during her morning inspection. She ignores what she sees on her return visit. If Baerends removed a full-grown larva and its cocoon from a cell, she nevertheless sealed the empty nest on her return visit. If he substituted a large larva for a small one, she still brought in only enough food for a small larva. If he packed a nest with more than enough caterpillars for the resident larva, she nevertheless stuffed in more caterpillars.

Social insects live in groups of from dozens to a million or more individuals that cooperate to raise the young of the colony. Among the more advanced social species, notably termites, ants, and honey bees, the individuals act in such harmony that their colonies have been called superorganisms. Proponents of the superorganism concept point out that the social functions of an insect society are integrated to operate as a whole just as are the functions of the different tissues and organs of an individual organism. In *The Insect Societies,* Edward O. Wilson of Harvard University wrote of the social insects: "together with man, hummingbirds and the bristlecone pine, they are among the great achievements of organic evolution. Their social organization—far less than man's because of the feeble intellect and absence of culture, of course, but far greater in respect to cohesion, caste specialization and individual altruism—is nonpareil."

The ultimate function of an insect society is, of course, the production of offspring that will found new colonies. All social insects give parental care, and they differ from nonsocial species in that fertile eggs are produced only by a specialized reproductive caste—usually only one queen per colony—but are cared for by a nonreproducing caste consisting of many workers. In all but the simple insect societies, the workers are kept sterile by a pheromone, produced by the queen, that prevents their ovaries from developing. But how does a caste that does not reproduce itself persist in evolutionary time? If our ideas about evolution are correct, the sterile workers must gain more fitness by helping their mothers produce offspring than they would by reproducing themselves. Our knowledge of honey bees, ants, and social wasps shows that this seeming paradox can be the truth of the matter. Assuming that a female mates only once, a peculiarity in the genetics of these insects determines that the workers (which are all females) are more closely related to their sisters (by three-quarters) than they would be to their own young (by only one-half). Clearly, in this case a worker could project more of her own genes into the future by helping a queen to raise sisters than she would by producing offspring of her own. Since queens usually mate with more than one male, workers may gain little or no advantage by helping the queen unless they can recognize eggs or larvae that were fertilized by sperm from their own father and thus concentrate on raising their own full sisters.

While colonies of termites, ants, and honey bees survive the winter in the North Temperate Zone, colonies of wasps and bumble bees do not. Each spring new wasp and bumble bee colonies are founded by queens that mated late the previous summer and survived the winter by diapausing in crevices, hollow trees, or other protected places. In Illinois, the queens emerge in April and May. Bumble bee queens then seek out a hole in the ground, often the abandoned burrow of a white-footed mouse, in which to found a colony. The lone queen begins by building a waxen cup into which she places a ball of pollen and honey. After she lays about a dozen eggs on the ball, she seals the cup by roofing it over with a dome of wax. While she waits for her young to develop, she fashions a waxen honey pot that she fills with nectar. After the first workers emerge, the queen continues to lay eggs, but the sterile workers, ultimately a hundred or more of them, take over all of the other duties of the colony. In late summer, a large number of queens and males are produced. They leave the colony and mate elsewhere. The males soon die, but the queens

A worker bumble bee
gathers pollen from a
black-eyed Susan

survive the winter as I have already described. The life cycles of
social wasps are similar, except that they build nests of paper and
feed their larvae insects.

⟍ Just as European cuckoos and North American cowbirds
lay their eggs in the nests of other birds, a few bumble bees have
become nest parasites of other, more industrious bumble bees.
There are no workers in these parasitic species; all of the females
are queens. These parasitic queens enter the nest of a nonparasitic
bumble bee and sting to death the rightful queen and sometimes
her larvae and pupae. Then they lay their eggs, which are cared
for by the surviving workers of the host colony.

An individual social insect survives and attains evolutionary fitness through membership in its society. No social insect could survive for long or reproduce itself if it were separated from its colony, except for overwintering wasp and bumble bee queens. Thus, much like an individual of any nonsocial species, a colony must feed itself and grow; it must protect itself from destruction; and it must reproduce itself. A brief look at honey bees, which have one of the most advanced societies known among insects— and insect societies are second in complexity only to human society—illustrates how evolution has solved these problems on a societal level.

In a hollow tree or in a wooden hive made by people, honey bees build hanging, vertical combs that are made of wax secreted by the workers. A comb consists of many hexagonal cells, in some of which the queen lays her eggs. Most of the cells are just the right size to accommodate a larval worker, but a few slightly larger ones are built for drone (male) larvae. When the time comes, a few much larger, pendant, free-form cells are built to house larvae destined to become queens. While many of the ordinary cells are used to rear larvae, collectively called "the brood" by beekeepers, the rest of them are either vacant or are used to store pollen or honey, the food of the adult workers themselves. The brood is fed a mixture of regurgitated sugars and the "bee's milk" that is secreted by young workers.

People and honey bees have had a long and, at least from the human point of view, a profitable association. An 8,000-year-old painting on the wall of a rock shelter in Spain shows a woman robbing honey from a colony of wild bees. The ancient Egyptians, Greeks, and Romans domesticated bees, and beekeeping is still an important industry. Hives are rented out to pollinate orchards and other crops, and the bees yield valuable products: honey, wax, royal jelly, and even venom used to desensitize people who are allergic to bee stings. Royal jelly is the special food that the workers feed to larvae that are destined to become queens. It is sold as a health food and as a promoter of virility for men, and is incorporated in cosmetics for women. In the last instance, the implication seems to be that if royal jelly can make a "super" female bee, it should somehow be able to enhance the femininity

of a woman. There is no scientific support for such a notion. There is also no evidence that royal jelly promotes virility in men or that it has nutritional qualities absent in other foods.

Royal jelly does not contain some wonderful substance that triggers a larva to become a queen rather than a mere worker. The "wonderful substance" is nothing more than sugar. Royal jelly is fed to queens in greater abundance and simply contains more sugar than the bee's milk fed to larvae destined to become workers. Hence, a larva destined to become a queen is stimulated by the sweetness of its food to eat more and grow larger. The truth of the matter is that any female larva that grows to a large enough size automatically produces, in her own body, a hormone that will cause her to develop into a queen. Although royal jelly is nutritious for humans, it does not contain a mysterious substance that enhances femininity. Growing up to be a queen is just a matter of eating more and getting bigger.

As we have seen, wasp and bumble bee colonies die off each autumn and must be started from scratch each spring by lone queens that survive the winter in diapause. But honey bee colonies survive the cold winters of our northernmost states and southern Canada without going into a diapause that would make them torpid and slow to respond. In the spring, the colony is intact except for a somewhat depleted population of workers. Just as soon as the first flowers appear, the workers are ready to collect pollen and nectar and to begin raising brood from eggs laid by the queen.

Although colonies of ants and termites also survive northern winters, honey bees are the only social insects that do so by generating heat to warm the colony. When the temperature gets down to about 55° F, the thousands of workers in a colony form a cluster around the queen. The workers at the periphery of the cluster form a tightly packed living blanket—two bees deep—that insulates the rest of the cluster. The inner bees are less densely packed and are free to shift around within the cluster. They are the source of the heat. They eat the honey stored in the combs and convert its calories to heat by vibrating their wing muscles without moving their wings. Throughout the winter, the bees usually manage to keep the temperature of the cluster between 68° F and 86° F when brood is present and somewhat lower if there is no brood. According to the *Illustrated Encyclopedia of Beekeeping*, one experimenter found that, on an unusually cold night, the temperature in the center of a cluster of bees was 106° F higher than the air temperature outside of the hive.

Honey bees also control the temperature in the hive during the summer, usually keeping it at about 95° F when brood is present. The brood survives and grows only within a narrow range of from about 90° F to about 97° F. When the weather is cool, the workers cluster more or less closely around the brood to keep it warm. But when the weather is too hot, they disperse about the combs and fan their wings to circulate fresh air through the hive. When air circulation alone is not enough to keep the hive cool, they bring in drops of water and evaporate them by fanning. This evaporative cooling works amazingly well. According to the above-mentioned source, when a hive was placed in direct sunlight in a location where the outside temperature was an astonishing 158° F, the bees managed to keep the temperature inside the hive at 97° F as long as water was available to them.

The economy of a honey bee colony depends upon the collection of an adequate supply of nectar and pollen, the food base for all of its members. In order to satisfy these needs, workers may forage as much as 7.4 miles from their colony, but 95 percent of the foraging occurs within a radius of 3.7 miles. Thus the workers regularly search an area of about 43 square miles for pollen and nectar. Worker bees are nicely adapted for this task. They can fly for long distances; their color vision and superb sense of smell enable them to locate flowers; each hind leg bears a basket-like structure that can be packed with a large load of pollen; and large quantities of nectar can be stored in the honey stomach and later regurgitated. But their most marvelous adaptation for the efficient collection of nectar and pollen is the symbolic language by which they communicate to each other the direction and distance of patches of blossoms that yield one or both of these substances.

A foraging scout that returns from a successful reconnaissance enters the hive, crawls directly to the vertical surface of a comb,

and performs a waggle dance that symbolically maps out the most direct flight path to the new source of nectar or pollen. The dance is in the form of a figure eight. As Karl von Frisch discovered, and as Thomas D. Seeley of Cornell University described in *Honey Bee Ecology*, the scout first makes a straight-line run, the cross bar of the eight, then turns to the right and circles back to the starting point and then makes another straight-line run and circles to the left to again return to the starting point. She repeats this pattern over and over. As she makes each straight-line run, she waggles her abdomen from side to side and emits a high-pitched buzz. In the darkness of the hive, other workers follow her as she dances, crowding in close and touching her with their antennae. These followers then fly to the newly discovered patch of flowers. If they are successful in finding pollen or nectar, they also dance when they return to the hive and thus recruit yet other workers that will visit the new resource. If the patch is highly productive, more and more workers will forage there and return to do the waggle dance. In this way, the attention of the colony is focused on a few productive patches of flowers.

The waggle run presents crucial information, the distance and direction of the newly discovered patch of flowers. Distance is indicated by the duration of the waggle phase of the dance—the greater the duration the more distant the flowers. The direction of the flowers in relation to the direction from the hive to the sun is indicated by the bearing of the waggle run of the dance with respect to the vertical axis of the comb. Just as we follow the convention that the top of a map is north, the bees understand that the top of the comb symbolizes the direction of the sun. If the direction of the flowers is directly away from the hive toward the sun, the worker does the waggle run straight upward along the comb. If the direction is directly away from the sun, she directs the waggle run straight downward. If the direction of the flowers is sixty degrees to the right of a line from the hive to the sun, the waggle run of the dance runs sixty degrees to the right of the vertical axis of the comb. The identity of the flowers is made clear by their scent, which permeates the pollen and the nectar that the bees bring back to the hive.

By 1923, Karl von Frisch, the central figure in the discovery and decoding of the waggle dance, for which he would later receive the Nobel Prize, had published a lengthy account of his early discoveries. In 1927, J. B. S. Haldane, inspired by von Frisch's work, wrote the following prescient words in his *Possible Worlds and Other Essays*:

> Tomorrow it looks as if we should be overhearing the conversation of bees, and the day after tomorrow joining in it. We may be able to tell our bees that there is a tin of treacle for them if they will fertilize those apple trees five minutes' fly to the south-east; Mr. Johnson's tree over the wall can wait! To do this we should presumably need a model bee to make the right movements, and perhaps the right noise and smell. It would probably not be a paying proposition, but there is no reason to regard it as an impossible one.

It was not until 1989 that humans first communicated with honey bees through the dancing and buzzing of a robot bee. Martin Lindauer, one of von Frisch's students, together with several other ethologists, devised a peppermint-scented robot—manipulated by a slender arm which protruded from the back side of a honeycomb—that danced so that the waggle run went straight up the comb. The experiment was a great success. Many bees from the hive appeared at a peppermint-scented bait placed at the distance and direction indicated by the robot. Very few appeared at other bait stations that were the same distance from the hive but in different directions. Robot bees may never be used to direct bees to plants that we want them to pollinate, but it is, nevertheless, quite an accomplishment for people to communicate with insects, and the use of robots will certainly help us to learn a great deal about honey bees.

Worker honey bees, the daughters of the reigning queen, are all sterile. Nevertheless, it is they who control all reproduction in the colony: the production of their brothers, the fertile drones, and the production of their sisters, a few fertile queens and thousands of sterile workers. Each year, a typical colony produces about 200,000 new bees. Every one of them develops from an

egg laid by the one and only queen. A few of these offspring, from 5,000 to 20,000, are drones, which develop only from unfertilized eggs. The other 180,000 or more are all females, which develop from fertilized eggs. Only about 10 of these females become queens; all of the rest are destined to become sterile workers.

Although the queen lays all of the eggs, it is the workers that control how many she lays and whether they become drones, workers, or queens. The queen lays only one egg per cell and will not lay at all unless there is an empty cell. Thus the workers determine how many eggs she will lay by increasing or decreasing the number of cells that they build. The queen lays unfertilized eggs that will become drones only in special cells that are somewhat larger than the cells in which workers are reared. The workers control the number of drones produced by limiting the number of drone cells that they build. Whether or not a fertilized egg becomes a worker or a queen is also determined by the workers. Fertilized eggs laid in small cells are fed ordinary bee's milk and become workers. Eggs laid in special, large queen cells are fed royal jelly and become queens. Workers can even change the destiny of eggs or small larvae in worker cells in the event of the death of the reigning queen. (Queenless colonies will eventually die unless a new queen can be produced.) If the queen dies and the colony has no queen cells with developing larvae, the workers build additions onto ordinary worker cells that contain an egg or a larva less than three days old. This larva is fed royal jelly and will mature to become a replacement queen.

There are two ways in which a colony of honey bees can send its genes out into the world and forward in time: through the drones that it produces or by dividing the colony into two parts and sending out one part, a queen accompanied by a swarm of workers, that will found a new colony.

On days with favorable weather, drones leave their home colonies to fly to aggregation sites where males from several colonies gather and to which new queens from several colonies orient in order to mate with the drones. This is similar to the lek formation

I discussed earlier. Both drones and queens usually fly to aggregation sites that are at least a mile away from their home colonies. Hence, brothers and sisters are not likely to end up at the same site and mate with each other. The drones fly about over the aggregation site until a queen arrives and releases a sex-attractant pheromone. She almost immediately acquires an entourage of drones that follows behind her as she flies on. When one of the drones catches her, the pair mate in flight. A drone mates only once. He dies after his genitalia explosively evert into the queen's vagina and tear loose from his body. The new queen generally mates with several drones before returning to her colony. She may return to mate several more times on a second or even a third day.

Successful colonies that have a large population of workers reproduce by swarming, dividing the colony into two groups, each headed by a queen. When the population of bees has grown so large that the nest is congested, the workers construct several queen cells. A few days before the first new queen is due to emerge from her cell, the old queen—accompanied by thousands of workers that have engorged on stored honey—leaves the nest to found a colony elsewhere. The swarm flies to a temporary resting site on a bush or a branch of a tree. There the bees hang in a tight cluster that surrounds the queen. Scout bees from the swarm go off in search of new nesting sites. Those that find a possible site, perhaps a hollow tree or a space in the walls of a building, return to perform on the surface of the cluster a waggle dance that indicates the direction and distance of the new nesting site. Scouts that have found high-quality sites dance more persistently than do those that have found low-quality sites. Workers in the cluster follow the scouts as they dance and then fly off to examine the new site. If they find it to be of high quality, they return to the swarm and repeat the same dance to recruit other workers. The swarm eventually arrives at a true consensus. When all of the dancing bees are performing the same waggle dance, almost invariably the one that indicates the best nesting site, the swarm flies off and moves into the new site.

If the home colony is still congested after the departure of the first swarm, a second or even a third swarm will leave with one of the new queens. After all of the swarms have left, the next new queen to emerge stings to death or otherwise kills all of the other developing queens in their cells. (If two queens happen to emerge at the same time, an unusual circumstance, they will fight to the death.) The surviving queen then assumes her royal role. When the colony grows large enough, it will again produce one or more swarms.

Defense against Predators

Insects are surrounded by predators. As plants are food for hundreds of thousands of species of grazing and browsing animals, ranging from insects to elephants, insects are forage for a host of predators that includes other insects, lizards, birds, mice, and even some large mammals such as bears. Denis Owen stated it nicely in *Camouflage and Mimicry* (1980): "No animal is safe. There are predators everywhere—waiting, lurking, running, jumping, burrowing, flying and swimming—in woods, fields, swamps, rivers and lakes; in the depths of the ocean and high on the mountains. If an animal is seen, heard or smelt it is potentially in danger."

The foremost vertebrate predators of insects are the birds that most people love and admire. We delight in their beautiful plumage; we marvel at their graceful flight; and we find joy in their songs. Great artists paint birds. Composers are inspired by their songs. Baseball teams are named for them. In her lyrical *Bird Neighbors*, first published in 1897, Neltje Blanchan quotes from poems extolling the beauty and virtue of birds by several American writers. James Russell Lowell wrote of the Baltimore oriole, now known as the northern oriole:

> Hush! 'tis he!
> My Oriole, my glance of summer fire,
> Is come at last; and ever on the watch,
> Twitches the pack-thread I had lightly wound
> About the bough to help his housekeeping.

Twitches and scouts by turns, blessing his luck,
Yet fearing me who laid it in his way.

Ralph Waldo Emerson wrote of the black-capped chickadee:

Piped a tiny voice near by,
Gay and polite, a cheerful cry—
Chick-chickadeedee! saucy note
Out of sound heart and merry throat,
As if it said, 'Good-day, good Sir!
Fine afternoon, old passenger!
Happy to meet you in these places
Where January brings few faces.

That is the human view of birds. I have been a compulsive birder for most of my life and I go along with this favorable attitude. I have no quarrel with any bird. But when we look at birds through the eyes of an insect, or some other tiny animal, the view is quite different. Birds are too small to eat people, but they are just the right size to eat insects, worms, and other small creatures. One of Gary Larson's "Far Side" cartoons gives us a worm's-eye view of birds: A group of earthworms is lined up in front of a theater to see a horror film. The title of the film is *The Robin.*

Insects and other animals have evolved various ways of avoiding predators or defending themselves against them. Many are camouflaged so that they are hard to find. Some animals frighten away predators by resembling an even more threatening predator, usually one that is much larger than they themselves are. This sounds impossible, but a large caterpillar can pass itself off as only the head and neck of a small snake, and a large moth with its wings spread can look like the face of an owl. Other animals sting, bite, or are otherwise obnoxious. A few are actually defenseless but bluff by resembling one of the obnoxious species. None of these tactics is infallible. They can only improve the odds for escape. Predators generally do find enough prey to keep themselves fed. Nevertheless, slightly better camouflage or a somewhat more painful sting does give a prey animal a small survival advantage over members of its own species that are not so fortunately endowed. No matter how small the advantage is, the better-protected individuals will eventually supplant the others.

Most animals are camouflaged. They blend in with their backgrounds. The dead leaf butterflies of India, true to their name, rest among dead leaves and are colored in matching shades of brown. A pattern of lines on the wings suggests the midrib and veins of a leaf. Certain fish of the Amazon Basin tend to escape their predators, and may gain an advantage over the small fish that are their own prey, by resembling a fallen leaf floating in the water. Their bodies are flattened sideways and are extremely thin. They float on their sides just below the surface and barely move their fins. Flounders, camouflaged fish that live in salt water, change their color and pattern to match the bottom sediments on which they lie. The expanded green wings of katydids look like the leaves among which they rest; some species even sport replicas of disease spots, insect nibbling, or other common blemishes. Nocturnal moths that rest on the bark of trees in the daytime are matching shades of gray and brown—except for a few white species with black markings that sit on the trunks of white-barked birches.

When another animal approaches, rabbits rely on their camouflaging coat of brown. Instead of running away, they freeze. They bolt only if the threatening animal comes too close for comfort. So it is with most camouflaged insects.

When they pick a place in which to rest, camouflaged animals choose a background appropriate to their appearance. Those that do not are soon eliminated by predators. A well-known German ethologist, Eberhard Curio, discovered that on the Galápagos Islands the young caterpillars of a hawk moth are always green, but that when they grow larger, some become either gray or brown although others remain green. The young green caterpillars and the older green ones sit on leaves when they feed and when they rest. The gray and brown ones also feed on leaves, but rest on twigs, whose bark they resemble. On Martinique, gray, brown and green anoles (small lizards) all hunt for insects together on the ground. When disturbed, they scatter and seem to disappear. The green ones run to green foliage, the brown ones to dry withered bushes, and the gray ones to the gray bark of trees. Many bark-resembling moths have dark streaks on their wings that mimic the crevices in bark. Different species assume different postures. Some sit with the body horizontal; others sit vertically. The wings may be held out to the side or they may be folded down over the body. But all of them sit so that the streaks on the wings are parallel to the crevices in the bark, thus giving the maximum camouflaging effect.

Camouflaged animals generally "blow their cover" if they move. But in a few special cases, appropriate movements enhance camouflage. As Hugh B. Cott reports in *Adaptive Coloration in Animals*, various leaflike butterflies, such as the dead leaf butterfly, may gently sway from side to side to simulate movement in the breeze. Bitterns, both the American and the Eurasian species, live among cattails or reeds. When alarmed, they stop and stretch their necks straight up into the air. Subtle stripes on the neck blend in with the vertical stems of the surrounding vegetation and make the bird almost impossible to see. If a breeze stirs the stems, the bittern sways from side to side in time. In *A Naturalist in the Guiana Forest*, Major R. W. G. Hingston, leader of the 1929 Oxford University expedition to British Guiana, described the camouflage-enhancing movements of an immature mantis. It sits head down on the bark of a tree. Most of its body is gray with patches of green and yellow that simulate lichens growing on the bark. But its abdomen, which droops down over its back, is gray-green and resembles a leaf. The mantis mimics the trembling of a leaf in the breeze by swaying only its abdomen, "sometimes gently, at other times vigorously, just as a leaf is made to move when touched at one time by a puff of air and at another by a distinct breeze."

In the summer of 1967, Aubrey G. Scarbrough, one of the graduate students who was then working with me and my colleague James G. Sternburg, had a curious experience that demonstrates the survival value of camouflage. Over a dozen abnormally colored, blue cecropia caterpillars appeared among the scores of normally colored, green caterpillars that hatched from eggs that he had confined under a net on a small apple tree. These are the only naturally occurring blue cecropia caterpillars that have ever been reported. Before the caterpillars could spin their cocoons, an unidentified vandal with a BB gun, probably a mischievous boy, invaded our research area and shot all but three of the conspicuous, blue caterpillars right through the net. None of the green caterpillars under the same net were shot. They apparently escaped his notice because they were so well camouflaged among the apple leaves. Birds rather than small boys are the usual agents of natural selection that weed out poorly camouflaged caterpillars, but, like small boys, birds hunt by vision and they see colors.

Insects of some species live on flowers and are camouflaged to match their colors. In much of North America, a large, yellow and black long-horned beetle, the locust borer, may be found on inflorescences of goldenrod, all but invisible among the mass of tiny yellow flowers. Sometimes an insect's resemblance to a flower goes beyond a mere blending in with its colors. Some planthoppers gather together in groups so as to enhance their resemblance to flowers. Each individual resembles a single blossom, but when they line up head to tail along a stem, all with their heads pointing in the same direction, they look like a spike of flowers, such as those borne by their food plants.

Natural selection is the core of Darwin's theory of evolution. He realized that it produces new species much as dog fanciers produce new breeds by artificial selection, by choosing only dogs with desirable, heritable traits to be the parents of the next generation. Similarly, natural selection tends to favor reproduction by those individuals that are best adapted to their environment. It tends to eliminate poorly adapted individuals and to permit the survival of individuals with characteristics that better enable them to avoid the hazards of their environment or to take advantage of the opportunities that it offers. For example, a well-camouflaged individual may be less likely to be eaten by a predator than is a poorly camouflaged one; a moth with a long tongue may be able to reach the nectar in flowers that are too deep for a moth of the same species with a somewhat shorter tongue. Individuals that successfully run the gauntlet of natural selection will be the parents of the next generation. Natural selection will eventually spread their useful new traits throughout the population. If two populations of the same species become separated geographically or perhaps by habitat or food plant, the accumulation and integration of new traits, especially courtship patterns that prevent mating between the two populations, will produce a new species: a population of organisms that "breeds true" and that is reproductively isolated from all other populations.

New, heritable characteristics constantly arise through the mutation of DNA, the genetic material that determines the form, the physiology, and the behavior of organisms. Mutations are caused by ultraviolet light, cosmic rays, emissions from radioactive substances, by certain chemicals, and, as molecular biologists have shown, by intrinsic factors in the genetic material itself. Natural selection tends to eliminate the majority of mutations, which are deleterious, and to preserve the few useful ones.

We usually think of natural selection as proceeding so slowly that evolutionary changes are imperceptible within a human lifetime. However, natural selection is going on now and has been observed in action several times. The most widely known and one of the best-understood cases is that of the beautifully camouflaged peppered moth—elucidated mainly by the work of H. B. D. Kettlewell of Oxford University. Before the Industrial Revolution, peppered moths were always or almost always light in color. They were inconspicuous on the light-colored bark of trees, and their camouflage was further enhanced because their wings and bodies were patterned so as to resemble lichens that grew on the bark.

But the forests in which these moths live changed after the Industrial Revolution began with the invention of the steam engine in the late eighteenth century. As coal-burning factories proliferated in the nineteenth century, smoke pollution increased. The trunks of trees in woodlands near factory towns were stained black and the lichens that grew on their bark died. The light-colored moths were no longer camouflaged when they sat on tree trunks. Only black peppered moths, which were not known to exist in the eighteenth century, would have been likely to escape notice on the smoke-stained bark.

The first black peppered moth, presumably a mutant, was found in a previously all-light population at Manchester in 1848. The better-camouflaged black form of this moth then increased as the population of the light form decreased. By 1898, only fifty years later, black peppered moths had almost supplanted light-colored ones in the vicinity of Manchester. By then, about 95 percent of the population consisted of black moths. The same thing happened in and around factory towns all over Britain, northern Europe, and, with other species of insects, in North America. Although some black mutants may have appeared in areas where there was no air pollution, natural selection did not favor them, and populations of peppered moths in clean areas continued to consist of light-colored individuals. The story of this moth has an interesting sequel. As Britain makes progress in cleaning up air pollution, the trunks of once stained trees are reverting to their original light color and the lichens are again growing on their bark. This favors the small minority of light-colored moths that survived in these areas, and they are now supplanting the black moths.

An important question remains unanswered, the identity of the selective agent. An obvious possibility is that birds that scan tree trunks for prey are most likely to find those peppered moths that do not match the color of the bark on which they rest. Kettlewell and his colleagues eventually showed this explanation to be correct, and others found that the black moths have the additional advantage of being better able than light moths to cope with the physiological stress that results from smoke pollution. Two tactics were then used to show that birds are important in the selection process. First, moths were released on both light and dark tree trunks and later recaptured at lights or traps baited with pheromone-releasing females. Moths released on bark of the "right" color were more often recaptured than those released on bark of the "wrong" color, which indicates that they were more likely to survive if they matched their backgrounds. The next step was to determine if birds are responsible. Moths were placed on match-

ing or contrasting tree trunks and watched from a blind. Birds were often seen to capture those that did not match the color of the bark on which they sat, but they very seldom noticed moths that did match their background.

　Certain insects escape the notice of predators not by blending in with the background but by resembling some obviously inedible and not necessarily inconspicuous object in their habitat—perhaps a thorn or a blossom. Some spiders and insects look like bird droppings. Among them are North American species, including caterpillars of some of our swallowtail butterflies. A bird-dropping caterpillar found in India was graphically described in a letter from Colonel A. Newnham to Professor E. B. Poulton of the University of Oxford:

> I came across the larva in question in the month of August or September 1892, at Ahmadabad on a bush of *Salvadora* . . . I was stretching across to collect a beetle and in withdrawing my hand nearly touched what I took to be the disgusting excreta of a crow. Then to my astonishment I saw it was a caterpillar half-hanging, half-lying limply down a leaf. [A] thing that struck me was the skill with which the colouring rendered the varying surfaces, the dried portion at the top, then the main portion, moist, viscid, soft, and the glistening globule at the end. A skilled artist working with all materials at his command could not have done it better.

Some moths resemble bird droppings with a different appearance. In addition to his previously cited description of a camouflaged mantis, Major Hingston described a small moth whose wings are glossy white, almost pearly, with patches of dark brown and areas with a blush of slaty gray and traces of yellow. It habitually rests on the upper side of a leaf with its wings spread out and flattened down against the surface. Major Hingston reported that when the moth sits in this position it looks exactly like a small bird dropping that fell from a height and flattened itself against the leaf. The caterpillars of an African moth also look like bird droppings and sit on the upper side of the leaves of their food plant. They are gregarious when they are young, and a group of them gives the impression of numerous small feces dropped by several small birds roosting above. When they grow large, the caterpillars become solitary and look like a single dropping of a large bird.

When camouflage or some other early line of defense fails, most animals, including many insects, actively defend themselves against predators. As mice bite when attacked or as herons stab with their sharp bills, beetles may use their jaws to bite, and robber flies and true bugs may stab attacking insect eaters with their piercing and sucking beaks. Grasshoppers and cockroaches kick out with powerful hind legs armed with spines. Skunks and bombardier beetles discharge sprays of noxious chemicals, and most female wasps and bees, as well as some snakes, can inject venom.

Honey bees, especially the race that originated in Africa, are famous for their painful stings and their belligerent defense of the colony against nest robbers such as skunks, bears, and humans. Drones are defenseless, but all females, including both workers and queens, are armed with a stinger. The virtually barbless stinger of the queen is not used in defense. Queens sting only other queens in battles that occasionally occur within a colony. The workers' stingers are barbed and, like those of the queen, are served by venom glands and a storage sac for venom. If the nest is disturbed by a large animal, thousands of workers rush forth to drive it away. (The same response can be provoked by banging on a hive with a brick.) The intruder may receive hundreds or even thousands of stings. Such mass attacks are usually fatal to humans. The barbed stingers of the workers remain embedded in the flesh of the intruder. When the worker tries to free herself, the stinger and venom sac tear loose from her body. The worker will eventually die, but the detached venom sac will for a short time continue to pump venom into the intruder's flesh.

Although all workers have stingers and may be involved in the defense of the nest, it was recently found that certain workers are much quicker to respond than others. In a 1990 article in *Behav-*

ioral Ecology and Sociobiology, Michael D. Breed, Gene E. Robinson, and Robert E. Page, Jr., reported that the first bees to attack an intruder are genetically different from workers that forage for pollen and nectar. Their wings are also less worn, which indicates that they fly less often than do foragers. These bees apparently constitute a previously unrecognized soldier caste—a "rapid deployment force" always ready to defend the nest.

Honey bees of the African race are good honey producers in the tropics, but they are notorious for their fierce defensiveness. They attack en masse with little or no provocation, and in South America have killed many people, many more domestic animals such as cattle, and no one knows how many wild animals. In 1956, with the approval of Brazilian federal and state authorities, Brazilian scientists imported a large number of African queens. Their intention was to produce a more productive variety of honey bee by hybridizing the African race with their own domestic bees, which were of the much less aggressive European race. In 1957, twenty-six African queens, accompanied by swarms of European workers, escaped from an experimental apiary near Rio Claro, which is about 100 miles north of the city of São Paulo. The escapees prospered and interbred with bees from feral colonies of domestic bees and even with bees from commercial hives. Soon virtually all of the feral bees in Brazil were "Africanized," and they quickly began to spread to the south and the north at a rate of from 100 to 200 miles per year. Africanized bees, often called killer bees, especially in horror films, now occupy all but the southernmost parts of South America and all of Central America and Mexico. There have been numerous reports of unprovoked attacks on humans in South and Central America, and hundreds of people have been killed by them. In the fall of 1990, Africanized bees reached the Rio Grande Valley in southern Texas. As I write this they are established in southern Texas, New Mexico, Arizona, and California, and there have been several attacks on people. Unfortunately, one of these attacks was fatal. The infested area has been quarantined to slow their northward spread. How far north they can spread is not known, but since they occur naturally in temperate areas of Africa and have become established in temperate areas of South America, it is possible that they will eventually occupy all or most of the southern United States. Mark L. Winston of Simon Fraser University in Burnaby, British Columbia, tells the story of the introduction and spread of the African bees in his recent book, *Killer Bees*.

Insects and other animals that sting, bite, or are otherwise obnoxious usually present warning displays that may ward off the predator before their weapons have to be used. The warnings are given in self-interest, not because of a kindly disposition. After all, it is risky to let things go so far that the defense of last resort has to be used. Before striking, cottonmouths gape to reveal the startlingly white linings of their mouths and rattlesnakes usually rattle. Wasps and bees give ample warning. They are strikingly colored, make loud buzzing noises, and otherwise behave so as to make themselves conspicuous.

Skunks present a warning display whose meaning is unmistakable. Their conspicuous patterns of contrasting black and white are in themselves warnings, glaringly obvious exceptions among the small mammals, most of which are inconspicuously camouflaged with browns and grays. When striped skunks, common in southern Canada and most of the United States, feel threatened, they fluff out and raise their tail in the air as straight as an exclamation point. If the threat continues, they turn their backside to the intruder and, tail still raised, drum out a rapid tattoo on the ground with their front feet. The spotted skunk, which occurs only west of the Mississippi River, fluffs out its white tail, does a headstand, and moves backward toward the intruder on its front feet, displaying the full length of the underside of its black and white body. If the intruder does not back off, both of these skunks spray their chemical weapons, which are not only smelly but also irritating to the eyes.

The chemical defense of the bombardier beetle is on a much smaller scale, but drop for drop it is probably more potent than are the defenses of skunks. In a 1972 English-language article in a German publication, Thomas Eisner of Cornell University described this beetle's armament. The beetle, which is warningly colored with bright blue and orange, has at the tip of its abdomen

a specialized organ that prepares and discharges the defensive fluid. This organ has two reservoirs, one containing hydrogen peroxide and the other highly deterrent chemicals called hydroquinones. If a threatening predator ignores the bombardier beetle's warning coloration, the beetle raises its backside and turns it toward the predator. At the same time, the contents of these two reservoirs are passed into a third chamber in which they undergo an explosive chemical reaction in the presence of certain enzymes, spraying the defensive chemicals, benzoquinones, toward the predator at the temperature of boiling water. Toads that try to eat these beetles are sprayed in the mouth and show obvious signs of distress, gaping wide and rubbing their tongues against the ground. In the meantime, the beetles make their escape.

The slow-moving and flightless lubber grasshopper of the southeastern United States has a more spectacular warning display than the bombardier beetle. While most grasshoppers are camouflaged when seen against their usual backgrounds, the large, heavy-bodied, yellow and black lubber is flagrantly conspicuous even in repose. At the sight of an intruder it raises itself on its legs and partly elevates its short but brilliantly crimson front wings. If it is touched by a pencil point, or presumably by the beak of a bird, it raises its front wings the rest of the way. This final display is accompanied by, followed by, or may even be preceded by the explosive discharge of a noxious substance that has a persistent odor somewhat like that of creosote or carbolic acid. Murray S. Blum, once my fellow graduate student at the University of Illinois and now a professor at the University of Georgia, recently told me that birds, reptiles, and some mammals are obviously frightened by this display and quickly back off. If a bird should happen to eat a lubber grasshopper, it becomes ill, vomits and will thereafter refuse to so much as touch one of these insects.

Other insects also contain poisons that do not affect a predator until after it has swallowed one of them. Although Jonah survived his encounter with the "great fish," being swallowed is almost always fatal. What purpose could be served by a poison that is effective only after the death of the animal that it is supposed to protect? The answer hinges on the fact that the predators, which are made ill by the poison but are rarely killed by it, soon learn to recognize the warning signals of these poisonous insects and reject them on sight.

The victim's death thus tends to protect other members of its own species. From the point of view of the rapidly growing science of sociobiology, if these other members of the species are relatives of the victim, perhaps its own offspring, sisters, or first cousins, their genetic makeup will be similar to the victim's. Thus many of the victim's genes will survive in these close relatives. In the case of a human, a bird, or almost any insect other than a bee or wasp, one-half of its genes will survive in any one of its offspring, one-half in a sister or brother, one-quarter in a nephew or niece, and one-eighth in a first cousin. Thus by sacrificing itself, the victim can enhance its own "inclusive fitness" by promoting the survival of some of its genes in its relatives. One way to understand this seeming paradox is to view an organism as the genes' way of surviving by making copies of themselves—something like the not altogether improbable argument that a chicken is just an egg's way of making another egg. In other words, an organism's fitness is ultimately a question of the survival of its genes.

Some insects manufacture poisons within their own bodies, but many others obtain poisons from their food plants. Monarch butterflies, whose black and orange wings are attention-demanding advertisements, contain toxic substances that they acquired from the milkweed plants that they ate when they were caterpillars. Called cardenolides, these toxins are related to digitalis and have therapeutic value for human hearts, but they can be deadly in rather low doses. Fortunately for an inexperienced bird that eats a monarch, these poisons cause vomiting at a slightly lower dose than the lethal dose and are thus usually eliminated before they can be fatal.

Lincoln Brower raised some monarchs on milkweed plants that do not contain these digitalis-like substances and other monarchs on milkweeds that do contain them. Blue jays that had been held in captivity long enough to forget any previous experience with a toxic monarch readily ate monarchs raised on poison-free plants.

These jays continued to eat nontoxic monarchs as long as they were offered. Jays that were offered monarchs that had been raised on toxin-containing plants ate them but quickly showed obvious signs of distress, erecting their crests and fluffing out their feathers. Then they became ill and vomited. After that, they refused to eat either toxic or nontoxic monarchs, and some of them retched when they so much as saw a monarch.

Poisonous insects are seldom deadly to their predators. From the point of view of the poisonous insects, it is more advantageous to make predators aware of their poisonous properties and allow them to live than it is to kill them. Most predators, including birds, are territorial and confine their activities to a small area from which they exclude, with the exception of their mates, other members of their own species. If the poisonous insect, the predator's victim, belonged to a species that does not wander widely, its survivors—sons, daughters, or other relatives—are likely to live in, or at least pass through, the territory of this predator that has already been educated not to attack them by the sacrifice of their relative. But if a territory-holding bird is killed, it is almost immediately replaced by a "floater," a member of its own species that is still searching for a territory of its own. If experienced predators are replaced by "floaters" that happen to be inexperienced, some of the insect victim's relatives may be attacked. Even animals that can educate predators without necessarily sacrificing themselves—wasps, bumble bees, or venomous snakes—may be forced to protect themselves again and again if a territory is reoccupied by a series of floaters. This is time-consuming and risky. A stinging insect—even one with a barbless sting—or even the most deadly snake may be injured or killed in an attack on its prey. Thus protective systems based on benign poisons are more likely to persist through evolutionary time than are systems based on deadly poisons.

A few animals that make predators ill are inconspicuous to the human eye, but most of them make themselves highly conspicuous, just as do skunks, bombardier beetles, and lubber grasshoppers. The warning coloration of noxious animals, often combinations of black with white, yellow, orange, or red, is often accompanied by other warning signals: loud sounds, twitching movements, wing wagging, or other behaviors that make them even more conspicuous. Like other bees and wasps, honey bees are warningly colored, in their case with alternating bands of yellowish-orange and black. Their warning colors probably do not deter skunks and bears that attack the nest, but they probably do warn off birds and other small predators that may attack lone, foraging bees away from the nest. The striking appearance of bees, wasps, and other brightly colored poisonous or venomous animals is readily learned by a predator that samples one of them for the first time, and thereafter their appearance serves as an easily understood and unmistakable warning that is visible from a distance.

The evolution of protected animals that emit warning signals paved the way for the subsequent evolution of animals that mimic them, harmless animals that bluff their way past predators by falsifying the warning signals of a truly protected animal. Henry W. Bates, a British naturalist, published the first account of this sort of mimicry, based on observations that he made during a long sojourn in the Amazon Valley of South America. It appeared in the *Transactions of the Linnaean Society of London* in 1862. This was only three years after the publication of Charles Darwin's *Origin of Species*. The debate between the Darwinists and the anti-Darwinists raged. Many opposed Darwin's theory of evolution on religious grounds, maintaining that life arose not through gradual evolution but by special creation. Some even argued that all living things were literally created in six days in the year 4004 B.C., a date calculated from biblical genealogies by Bishop James Ussher of the Anglican Church of Ireland in the seventeenth century. Bishop Ussher reckoned the date and hour of the creation to be October 23 at noon. At times the discussions were acrimonious. During one debate, the Anglican prelate Samuel Wilberforce asked Thomas H. Huxley, Darwin's foremost defender, if he was descended from an ape on his mother's side or on his father's side. Huxley won that debate, probably in part because of Wilberforce's ill-considered remark. In this atmosphere, the pro-Darwinists welcomed Bates's account of mimicry as an obvious and dramatic example of the evolutionary process.

Batesian mimicry, as it is now called in honor of its discoverer, is most widespread among insects, but it also occurs among vertebrates. Burrowing owls and rattlesnakes both live in abandoned burrows in prairie dog colonies, and the snakes serve as acoustic models for the mimetic owls. When the owls are in a burrow, where they cannot be seen, they hiss at intruding predators, making a sound similar to the buzz of a rattlesnake. But when they are above ground, where it is obvious that they are not snakes, they scream and chatter at threatening predators as do other birds. Deadly coral snakes, as F. H. Pough has pointed out, are said to be models for milk snakes, scarlet king snakes, and other harmless snakes that mimic their black, red, and yellow color pattern. The resemblance is so close that an expert on snakes once named and described a harmless mimic as a new kind of venomous coral snake. The deadliness of the coral snake's venom seems to defy the proposition that protected animals stand to profit more if they educate rather than kill their predators. Furthermore, how can other snakes possibly benefit from mimicking a model that kills its enemies rather than educating them to avoid it? But at least some predators do not have to learn to shun coral snakes; they are born with an aversion to the coral snake's color pattern. It may even be that the coral snake is a mimic rather than the model. Coral snakes are resembled not only by completely harmless snakes but by nonlethal, venomous snakes as well. Some biologists have argued that these mildly venomous snakes, which sicken but do not kill animals that they bite, are actually the models for both the deadly coral snakes and the harmless snakes that resemble them. As would be expected if they are indeed mimics, coral snakes are reluctant to bite in defense, as snake collectors will verify. Plate 52 of John James Audubon's elephant folio shows two chuck-will's-widows on a branch with a coral snake. Audubon, apparently not knowing the venomous properties of this snake, handled and posed his potentially deadly model but was obviously not bitten by it.

There are many examples of Batesian mimicry among the

insects. In the Philippine Islands, certain harmless cockroaches are difficult to distinguish from inedible, red and black ladybird beetles; others closely resemble unpalatable, yellow and black leaf beetles. The females of one species of African swallowtail butterfly come in several different color forms that resemble different noxious butterflies that differ greatly from each other in appearance. A well-protected, tropical relative of the North American bombardier beetle is convincingly mimicked by a harmless grasshopper. The edible drone fly mimics honey bees. A hawk moth resembles bumble bees, and certain other small day-flying moths are mimics of wasps. Unlike other moths, those that mimic wasps or bees have transparent wings like their models. Soon after emerging from the pupal case, they shed the covering of scales that makes the wings of other moths and butterflies opaque.

A large North American hover fly (*Spilomyia hamifera*), which has no common name, is a frighteningly accurate Batesian mimic of stinging yellowjacket wasps. So convincing is the resemblance that these flies occasionally deceive even entomology students. I have found *Spilomyia* pinned among the wasps in their insect collections. But the two are very different insects and are not related. The fly has one pair of wings, while the wasps have two pairs. The wasps are social and raise their young on insect prey in communal underground nests. If their nest is disturbed, they swarm out to sting the intruder. *Spilomyia*, by contrast, is nonsocial, and its maggots live in and feed on the debris in wet, rotting cavities in trees. It visits flowers for nectar, as do the wasps, but it cannot sting.

This fly not only has a convincingly wasplike color pattern, but, as I argued in *Evolutionary Biology*, it also mimics several other prominent features of wasps. Its waist is somewhat narrowed, as is the waist of a wasp. Like most other flies, *Spilomyia* has short antennae that are barely visible to the naked eye. It mimics the long, black, and highly mobile antennae of wasps by waving its black anterior legs in front of its head. When the wasps sit on flowers, they fold and hold their lightly tinted wings out to the side. Their wings, when folded lengthwise in several layers, look like dark brown bands. The fly, too, holds its wings out to the side,

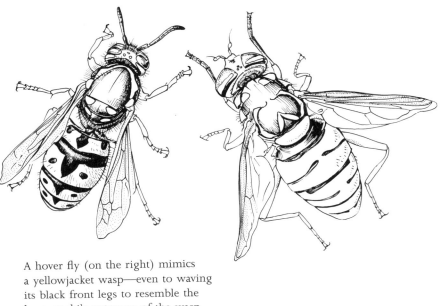

A hover fly (on the right) mimics
a yellowjacket wasp—even to waving
its black front legs to resemble the
long, mobile antennae of the wasp

but it cannot fold them. It mimics the appearance of a wasp's folded wing with a band of dark brown pigment that runs the length of the leading edge of the otherwise transparent wing. When they are on flowers, the wasps call attention to themselves by rocking from side to side. The flies do not rock, but they mimic this motion by wagging their wings. Finally, if the fly is grasped in the fingers, or presumably by the beak of a bird, it makes a loud sound that is almost identical to the squawk of a disturbed wasp.

Some biologists once thought that Batesian mimics and their models must occur together in the same place and at the same time. There is little evidence to show that mimics and their models can be separated in place, but it is theoretically possible; a northern insect might mimic a tropical insect if both are exposed to predators that migrate between the tropics and the north. There is, however, abundant evidence that a mimic need not coexist with its model in time. My colleagues and I found that *Spilomyia* and several other hover flies that mimic bumble bees or wasps occur only in early spring, when the flowers from which they obtain nectar are abundant but their models are almost absent. The models are not common until midsummer, weeks after the mimics have disappeared. We reasoned that the mimics are, nevertheless, protected because in the early spring before most birds have fledged their young, they are exposed mostly to adult birds that remember painful experiences they had with wasps and bees in a previous summer. One of my former graduate students, David L. Evans, compared the responses of adult and recently fledged birds to another hover fly, a species of *Mallota* that mimics bumble bees. He trapped newly returned, adult, migrant red-winged blackbirds and common grackles in the spring, when bumble bees were still scarce. When these newly returned, adult birds were offered mimics of bumble bees, they rejected them on sight, although the mimics are harmless and perfectly edible. Presumably, they remembered their experiences with real bumble bees from the previous summer. He also hand-raised nestlings of these birds in his home, getting up at all hours of the night to feed them and to make sure that they were warm. Evans tested the hand-raised birds when they had grown to adulthood. All of these young birds, which had never before seen a bumble bee or a mimic of a bumble bee, ate all of the mimics that he presented to them. That is, until they had been stung by a real bumble bee. After that they rejected mimics of bumble bees on sight.

Pipe vine swallowtails are large butterflies that occur in southernmost Ontario and much of the United States except for the northern tier of states. They are extremely noxious to birds because they store poisons that they obtain from the plants that they eat when they are caterpillars. Their only food plants are members of the birthwort family, noted for the toxins that they contain. In the midwest, pipe vine swallowtails feed on the native Virginia snakeroot, but they will also eat cultivated birthworts such as the ornamental vine called Dutchman's pipe. The warning coloration of these butterflies is striking. On their upper surfaces, the wings are largely black with a brilliant blue sheen. The under surfaces are black, with an iridescent blue area and large orange spots on the hind wings.

Several familiar and otherwise unprotected North American butterflies mimic the pipe vine swallowtail's appearance more or less faithfully: the female black swallowtail, the female diana fritillary, the black color phase of the female tiger swallowtail, and both sexes of the spicebush swallowtail and the red-spotted purple. The day-flying males of the promethea moth, closely related to cecropia, also mimic the pipe vine swallowtail, but only on the upper surfaces of their wings.

Why do only the females of some of these species mimic the noxious pipe vine swallowtail? Why aren't the males of these species protected against predators by mimicry? No one knows, but there are two possible answers. One is that females gain more from being mimetic than do males, because they are more exposed to predators, perhaps because they are more active. There is little supporting evidence from butterflies, but the sex-limited mimicry of promethea—to be discussed shortly—suggests that this hypothesis may well be correct. The other possible answer is that female butterflies are more likely to accept courting males with the ancestral color pattern of their own species than males with the more recently evolved mimetic pattern. The idea is that any gain in fitness that might accrue to the male through mimicry would be more than offset by his decreased ability to attract and inseminate females, but there is little evidence to support this hypothesis.

Female tiger swallowtails occur in two color phases: non-mimetic, yellow and black individuals that resemble the males and mimetic, black individuals that resemble pipe vine swallowtails. In the southern United States, where pipe vine swallowtails are abundant, almost all of the female tiger swallowtails are mimetic, except for those in an area in southern Florida where pipe vine swallowtails are absent. North of the model's range, they are all nonmimetic. (The northern tiger swallowtails are now considered to be a species different from the southern ones.) Where pipe vine swallowtails are present but not abundant, as in the upper midwest, both color phases are present.

Someone who has not closely watched tiger swallowtails in the field might expect the yellow and black males and females to be conspicuous and probably warningly colored. But their yellow and black markings are arranged in such a way that they obscure, or disrupt, the butterfly's form. When they sit on foliage they blend in quite well because of the disruptive pattern and because the yellow areas of the wings reflect green light from the leaves. Yellow and black bees and wasps, not disruptively patterned, are often conspicuous at a distance, but yellow and black tiger swallowtails are not necessarily so.

It makes sense that the mimetic color phase does not occur where there are no models to educate the predators, but why should there be two color phases of females at all? One possible answer is the intuitively pleasing notion that males prefer females with the ancestral yellow and black pattern. Where the model is abundant, the advantage of being mimetic might outweigh the advantage of being more attractive to males. Where the model is absent, there is no advantage to be gained by being mimetic, and only the nonmimetic color phase occurs. One researcher collected both mimetic and nonmimetic female tiger swallowtails in an area where both occur. Upon dissecting these females, he found that the yellow and black, nonmimetic females had been more frequently inseminated than had been the black, mimetic females, but his data are scanty and not statistically significant. Other researchers have since found that females of the two color phases are inseminated about equally often.

The sex-limited mimicry of the promethea moth is reversed. In this species, the males mimic the pipe vine swallowtail—at least on the upper surfaces of the wings, which are mostly black. The light orange-brown and creamy-white females are nonmimetic. Contrary to the situation with the butterflies, the mimetic sex, the males, are more exposed to diurnal predators than are the females, and they thus stand to gain more from being mimetic than do the females. The male promethea differs from the cecropia, and other closely related moths, not only in being mimetic but also in being a day-flier. Female promethea moths, however, fly only in the dark, when they lay their eggs. Like the female cecropia, the female promethea emits a sex-attractant pheromone, but she releases it only during the afternoon, when

she is hidden in foliage. The male must fly to her in broad daylight and thus expose himself to diurnal, insectivorous birds.

Jim Sternburg, Michael R. Jeffords, and I took advantage of promethea's unusual behavior to test the theory of Batesian mimicry under natural conditions. Promethea males are apt subjects for this experiment because they fly like butterflies, look like butterflies, and are active during the day like butterflies. But unlike butterflies, which must be recaptured one by one in hand nets, promethea males can be recaptured in large numbers, and almost effortlessly, in traps baited with pheromone-releasing females. There have been many demonstrations of the effectiveness of Batesian mimicry with captive birds in the laboratory. But there had been only one previous attempt, only partly successful, to demonstrate Batesian mimicry under natural conditions in the field.

Jeffords painted the largely black upper wing surfaces of palatable male prometheas with three different patterns. Some had their black wings painted only with black. This did not change their appearance—they still resembled pipe vine swallowtails—but it did serve as a control for the effect of painting the wings. (Although promethea males lack the blue sheen of pipe vine swallowtails, the two species are similar in appearance under some conditions because the blue sheen shows only when the light hits the swallowtail's wings at a certain angle.) Other prometheas were painted with yellow stripes so as to create a caricature of the nonmimetic, yellow and black, disruptive pattern of the palatable tiger swallowtail. Others were painted with orange bars to resemble the toxic monarch. Equal numbers from each group of painted moths were released in a natural area in the center of a mile-wide circle of seven traps baited with virgin, female promethea moths. Large numbers of the painted promethea males were recaptured in the traps, but only after they had run the gauntlet of insectivorous birds as they flew at least half a mile through mixed woodland and prairie from the center of the circle to one of the traps.

The results show that mimics really are protected by their warning signals. Moths painted to resemble poisonous pipe vine swallowtails or monarchs more often survived to be recaptured than did moths painted to resemble the edible tiger swallowtail. It also appeared that it was birds that distinguished between painted moths that looked like toxic butterflies and those that were painted to look like edible butterflies. Surviving yellow-painted moths often bore wounds that showed that they had escaped from an attacking bird. Large pieces of their wings had been torn loose, and on the intact wings of a few others, the triangular imprint of a bird's beak was clearly visible. Signs of an attack by a bird were much less often seen on the wings of moths that had been painted to resemble either toxic pipe vine swallowtails or monarchs.

The monarch and the viceroy butterflies are not closely related, but they are so similar in appearance that they are difficult to tell apart. The viceroy, long thought to be relatively palatable and protected only by its resemblance to the monarch, was until recently considered to be a classic example of a Batesian mimic. But the viceroy is actually unpalatable, although it is much less unpalatable than the monarch. It thus does not strictly fit the definition of a Batesian mimic, an edible animal that is bluffing. It is, rather, an example of Müllerian mimicry. In 1879, Fritz Müller argued that two or more poisonous and warningly colored species may come to resemble each other, because they will tend to evolve similar warning signals. Müller pointed out that no matter how toxic they are, some individuals will always be killed in the process of educating and reeducating predators. If two species have similar warning signals, there will be an economy for the members of both species, because the mortality will be shared among a larger number of individuals. Müllerian mimicry is widespread. Take the stinging wasps as an example. There are hundreds of different species, but most of them share the same "advertising logo." They are barred with yellow and black and make similar buzzing noises.

Blind as a bat is a false simile. But a bat's ability to see is not of much use as it flies through the dark hunting for insects. Bats use echolocation, or sonar, to locate their prey. Sonar is

familiar from old movies about submarines—the tense scene in which the submerged vessel hides from a destroyer on the surface. A slow ping, ping, ping, is heard—the destroyer's sonar, sound waves that it sends into the depths to probe for the sub. Echoes that bounce off the sub reveal its depth and its location. Bats discovered sonar millions of years before people did. As a bat flies through the night it gives off pulses of ultrasound, sounds that are too high-pitched for the human ear to hear. Echoes of these pulses bounce back from flying insects and twigs and other obstacles. The bat's hearing is so acute—many bats have extremely large ears—that it can hear these faint echoes and use them to avoid an obstacle or to locate its prey.

Long ago it was discovered that some, but not all, moths have ears, although they do not make sounds. What, then, do these moths listen to if they are not listening to each other? The answer is that they are listening for moth-eating bats, a phenomenon investigated and described in detail by the late Kenneth D. Roeder, who was an insect physiologist at Tufts University in Medford, Massachusetts. (See the article by M. Brock Fenton and James Fullard for more detail.) He noticed that some moths take evasive action when bats are nearby or even a considerable distance away. He attached fine wires from recording devices to nerves in the ears of restrained moths. In this way he himself could hear the bats' sounds through the ears of the moth.

When he played recorded bat sounds to moths that were swarming around a light outdoors at night, some of them folded their wings and fell to the ground, some made power dives or took other evasive action, and others, presumably the ones without ears, did not respond at all. Roeder's observations leave little doubt that moths with ears are better able to escape bats than are moths without ears, because they can hear ultrasonic cries, sometimes from a distance, and take appropriate action to avoid the hunting bats. You can try a simple experiment to convince yourself that some moths can hear. Jingle a bunch of keys near moths flying around a light. The jingling keys produce enough ultrasound to stimulate the moths with ears to take evasive action.

Owlet moths (noctuids) have a pair of ears, one on each side of the thorax just below the hind wing, but at one time no hawk moth was known to have ears. On a warm Texas night, Roeder was on a patio with other guests at a party. At the edge of the patio nocturnal hawk moths hovered in front of flowers as they sipped nectar. One of the guests asked Roeder if hawk moths have ears. He said that he did not know but took a bunch of keys out of his pocket and jingled them near the moths. The moths responded by taking sudden evasive action. This was the first evidence that any hawk moth has ears. When Roeder returned to his laboratory at Tufts, he looked for the ears of this species of hawk moth. They were not on the thorax below the hind wing, but he finally found them on two structures near the mouth.

Green lacewings, which are not related to moths, also have ears, one on each forewing. Lacewings respond to ultrasound by diving to the ground. Just as Cyclops had a single eye, some praying mantises have a single ear between their hind legs. These mantises abruptly alter their flight path when they are stimulated with ultrasound. The differing locations and differing structures of the ears on these various insects indicate that the ability to listen for bats evolved independently several times.

Some members of the tiger moth family not only have ears but also are able to produce ultrasound. Roeder and a colleague showed that these moths respond to the sounds of bats by producing their own sounds at about the same frequency as the echolocating cries of the bats. Some think that these sounds are attempts to "jam" the bat's sonar, and there is some evidence that this might possibly be so. There is also evidence that at least some tiger moths use these sounds to communicate with each other during courtship. However, it is likely that the sound signal also warns the bats of the moth's unpalatability. Bats do find tiger moths to be noxious. One of my friends, a fellow entomologist, kept tame bats in his home and gave them the freedom to roam where they would. They make good pets and seem to be quite intelligent. He had trained his bats to respond to ultrasound, which he produced by sharply rubbing his thumb and index finger together. The bats would fly to his hand, and he would reward them with an insect.

One night while I was watching, he called a bat from an upstairs bedroom down to the living room on the floor below. When he offered this bat a tiger moth, it refused to eat it and showed obvious signs of distress. He repeated this simple experiment many times with different bats, always with the same result. Although they are nocturnal, many tiger moths are brilliantly colored to warn off diurnal predators. Many of these moths spend the daylight hours resting on the upper side of a leaf where they are in plain view but protected by their warning colors, often stark white or combinations of black with orange or red.

The story of bats and moths is complicated by tiny, parasitic mites that live, feed and reproduce only in the ears of moths. Some of these mites live in the outer ear and do not impair the moth's ability to hear. They commonly infest both ears of a moth. Other mites, first discovered in 1954 by Asher E. Treat in Tyringham, Massachusetts, may infest either the right or left ear, but they are almost never found in both ears. These mites burrow deep into the inner ear, completely destroying its capacity to perceive sound. When one of these mites boards an uninfested moth, usually from a flower that had previously been visited by an infested moth, it shows considerable hesitation in deciding which ear to enter. But other mites that later board this same moth make their way almost directly to the ear occupied by the first mite, following almost exactly the same path that the first mite followed. There is not much doubt that the first mite laid down a pheromone trail leading to the ear of its choice.

A one-eared moth is handicapped in escaping from bats, but a moth with no functional ears at all is helpless. Both the moth and its parasitic mites are more likely to survive if all of the mites congregate in the same ear. To paraphrase Asher Treat, a moth that has been deafened in only one ear is a safer vehicle for mites than is a moth that has been deafened in both ears.

For creatures that eat plants, camouflage is mainly, if not wholly, a defense against predators. But camouflage probably serves small carnivores and insect eaters in two ways: as a means of deceiving their own prey, and to help keep them from becoming food for a larger and more ferocious carnivore. The camouflaging patterns of large carnivores such as tigers and jaguars probably function only as an aid in stalking prey. After all, the giant cats are the top predators in their ecosystems. There are no larger animals that hunt them. But small predators, especially insect eaters such as spiders and praying mantises, are almost as likely to become a meal as they are to catch one.

Our North American mantises are well camouflaged. Green mantises ambush their prey from their resting places among leaves. Brown mantises lie in wait on bark that matches their color. On the Malay Peninsula, however, there is a mantis whose camouflage is even more deceptive. This bright pink mantis rests among the pink blossoms of a plant known as the *sendudok* in the Malay language. Most of the mantis' body is pink, and on its middle and hind legs there are wide, thin flanges that resemble the pink petals of the blossoms of the *sendudok*. It is so deceptive in appearance that flies sometimes land on its body rather than the surrounding flowers. In a 1900 report of a University of Cambridge expedition to the Malay Peninsula, Nelson Annandale described this mantis and its behavior and told of some Malay beliefs about it. At Aring, a village in southern Siam (now Thailand), the natives agreed that the *kanchong*, the Malay name for this mantis, is

> but a flower which has become alive. "Its origin is from the flowers." The blossoms of the "*Sendudok*" give birth to it in the same way as the leaves of the "*Nanka*," or Jack Fruit tree . . . give birth to . . . a large prickly [stick insect] of great rarity, which rich men keep alive in cages in order to secure its [bright red] eggs, which they set in rings like jewels, and consider to be a most powerful charm against evil spirits of all kinds.

Crab spiders, predators common in Europe and North America, do not spin webs but wait in ambush to seize and suck their prey dry. Some species wait on blossoms to grab bees, flies, or butterflies that come in for nectar or pollen. Several of these species can change color. One of them can change from pink to yellow and back again, and another can switch between yellow and white. In The World of Spiders, the British arachnologist W. S. Bristowe described a simple experiment that gives strong support to the proposition that the crab spider's camouflage is an aid in catching prey.

He arranged sixteen yellow dandelion blossoms on a lawn, and in the center of each one he placed a pebble of about the same size as a crab spider. Half of the pebbles were black and highly conspicuous, and the other half were very inconspicuous because they were about the same color as a crab spider and matched the yellow of the dandelion. For half an hour Bristowe watched these dandelion heads—with the astonishing result that the ones with black pebbles were visited by only seven insects, while the ones with yellow pebbles were visited by fifty-six bees and flies. Bristowe said that this experiment satisfied him that crab spiders gain aggressive advantage from matching the colors of the flowers on which they sit and thus not alerting potential prey insects. He admitted that it is more difficult to prove that the resemblance protects them from their own predators. But he pointed out that among 10,000 spiders extracted from the stomachs of birds by the United States Biological Survey, only 4 were crab spiders.

There is a third type of mimicry that is known as aggressive mimicry. I have already mentioned one example, the firefly femmes fatales that devour the males of other firefly species that they lure with a falsified flash signal. Other examples of such treacherous impersonation are known throughout the Animal Kingdom.

When a giant alligator snapping turtle lies partially hidden in the mud at the bottom of the Mississippi or one of its tributaries, jaws agape and wriggling its wormlike red tongue, it is quite obviously using aggressive mimicry to attract a fish that will become its next meal. An even more bizarre form of aggressive mimicry is seen in some North American freshwater clams whose larvae are parasites on fish. The larvae attach themselves to the gills or fins of fish and remain there until they develop into tiny clams that fall to the bottom and burrow into the sediments. A full-grown female clam, which can be larger than a human hand, may discharge as many as 300,000 of these parasitic larvae at one time. Clams that live in still water release their larvae with little or no attempt to attract a fish. These larvae will lie on the bottom until a fish comes along and inadvertently sucks them up as it gulps a bit of food from the bottom.

Clams that live in streams where the current will wash their larvae away use a remarkable form of aggressive mimicry to attract large fish. The shell of the clam sticks above the bottom, and from a gap between the valves of the shell there protrudes a fleshy portion of the clam that is a convincing mimic of a small fish. Only the females have this lure fully developed. The little imitation fish has a tail, fins, and two eyelike spots on its head. The lure faces upstream and it twitches and undulates like a real fish. When a hungry carnivorous fish approaches the lure, the female clam releases a puff of parasitic larvae into its face. Some of these larvae are sucked into the fish's gills, where they attach themselves and begin their existence as parasites.

Some insects are the victims of aggressive mimicry, while others—and some spiders, too—are perpetrators of aggressive mimicry. Whenever insects are the victims, the modality of the falsified signal is more than likely to be one of the chemical senses, either taste or smell. Few insects have acute vision and many of them are deaf. Consequently, almost all insects depend largely upon their chemical senses, just as we rely mainly on vision and hearing. This is one of the main reasons why humans find it difficult to interpret and understand the behavior of insects.

Take, for example, bolas spiders, nocturnal predators that capture only night-flying moths. Mark K. Stowe of Harvard University, an expert on these remarkable spiders, reported that they attract moths from a distance, that these moths are all males, that they approach the spider from downwind, and that they can be caught in traps baited with bolas spiders. This is, of course, strong evidence that these spiders attract their prey by mimicking the odor of the sex-attractant pheromones of female moths. In a 1987 article in *Science*, Stowe and his colleagues provided conclusive evidence that hunting bolas spiders do indeed synthesize and release several compounds that are chemically identical to components of the sex-attractant pheromones of the moths on which they prey. The manner in which these spiders ultimately capture the male moths is unique among animals—except that humans have invented somewhat similar techniques, the bolas and the lasso. When a bolas spider is ready to hunt, it spins a few strands of silk by which it hangs from the surrounding vegetation. From one of its legs there dangles a "bolas," a separate length of silk that has a droplet of very sticky glue at its end. When a moth comes close enough, the spider flicks the "bolas" at it. With a bit of luck, the moth is held fast by the glue and the spider reels it in.

In 1911, Edward Jacobson of Samarang, Java, then in the colony of the Netherlands East Indies, published in a Dutch journal, Tijdschrift voor Entomologie, a fascinating account of how a Javanese assassin bug attracts and subdues the ants that are its only prey. The ant in question is a lover of sweets. Much as dairies keep cows, these ants maintain large herds of aphids, scale insects, or other excretors of honeydew on trees and shrubs near their colonies. (Honeydew, to be discussed in a later chapter, is the sugary and fluid excreta of aphids and related insects.) The predaceous caterpillars of a butterfly related to the hairstreaks, which I will say more about below, live among and eat the members of these insect herds. Although the ants destroy other creatures that prey on their "cows," they do not destroy these caterpillars. The caterpillars bribe the ants by secreting a fluid—probably containing sugars—that the ants like to drink. (It might also be that the caterpillar mimics a recognition pheromone that allows it to pass as an ant or a member of the herd.)

But back to the predaceous assassin bug. When one of these bugs is ready to eat, it stations itself along a trail that leads from the ants' colony to one of their herds. (It may recognize the trail pheromone secreted by the ants, but this is only my guess.) The ants are attracted to the bugs, possibly because the bugs mimic the appropriate pheromone of an ant or herd member. When ants approach, the assassin bug rears up so that it exposes an area on its underside that secretes a substance, presumably sugary, that the ants like to eat. The bug does not immediately kill the ant, but allows it to feed until a powerful tranquilizer in the bug's sugary secretion causes the ant to become paralyzed. Only then does the assassin bug suck the ant's body dry and let it fall to the ground. Piles of dead ants accumulate under the bugs. The corpses are carried away, presumably to be eaten, by an altogether different species of ant.

Even plants use mimicry to victimize insects. Some plants trick carrion-feeding flies into pollinating them. Their blossoms, which smell like rotten meat, attract the flies but supply no food for them. The bee and wasp orchids of Europe practice an even crueler version of aggressive mimicry. Their blossoms are readily recognizable but only roughly realistic mimics of their namesakes. The blossoms also give off scents that mimic the sex-attractant pheromones of wasps or bees, which is demonstrated by the fact that blossoms hidden from view in cages still attract male wasps and bees. Since the blossoms look, feel, and smell like females, the males may spend a long time making futile attempts to copulate with them. A duped male rarely places semen on the orchids and eventually becomes discouraged and flies away, but not before he acquires two bundles of pollen that stick to the front of his head and will fertilize the next orchid blossom that deceives the unfortunate male.

Certain insects and other animals gain a measure of protection by startling potential predators. Some nocturnal underwing moths, for example, spend the day resting motionless on the bark of a tree, their superbly camouflaged front wings covering their brightly colored hind wings. Humans, and presumably birds as well, usually fail to notice them when they rest in this position. But when one of these moths is disturbed by a person—and most likely also when it is disturbed by a bird—it lifts its inconspicuous front wings to reveal hind wings that are boldly striped with black and red, yellow, or white; the color is different in different kinds of underwing moths.

This startling display may serve the moth in two ways. First, the bird may be frightened off or at least taken aback, which will

A blue jay about to attack
an io moth is startled when
the moth lifts its front
wings to reveal large eyespots on its hind wings

give the moth precious seconds in which to escape. Second, if the
bird pursues the moth, it may be deceived into not noticing the
moth after it lands on the trunk of another tree. The pursuing bird
is focused on the bright, flashing colors of the hind wings,
strikingly visible as the moth flies off. But these colors suddenly
disappear when the moth lands and again covers its hind wings
with its front wings. The bird, which has a "search image" for
something brightly colored, is not likely to notice the now inconspicuous moth.

Startle responses are sometimes greatly enhanced by the presence of large, eyelike spots on the wings of insects. The io moth, a member of the same family as cecropia, has inconspicuous front wings that cover conspicuously banded hind wings that have in their center a large and realistic eyespot, complete with a highlight on the pupil. When the moth is disturbed, it lifts its front wings, suddenly revealing the staring eyespots. The moth seems to vanish and be replaced by the face of a glowering owl. It does not stretch the imagination to think that an attacking bird might be frightened by this startling apparition.

Convincing evidence that startle displays and large eyespots may indeed frighten birds was uncovered by A. D. Blest when he was a graduate student working on his doctorate at the University of Oxford in England. He showed that the sudden revelation of the eyespots on the hind wings of the peacock butterfly actually does evoke escape responses in birds. Butterflies with their eyespots intact often startled attacking yellow buntings. Butterflies from which the eyespots had been rubbed away seldom evoked a startle response. Blest did a different experiment with chaffinches, great tits, yellow buntings, and reed buntings. He placed a mealworm, a beetle larva that is a delicious morsel for any one of these birds, on a horizontal screen onto which he could project various images. When a bird was about to pick up a mealworm, he projected an image on either side of the insect. The birds were startled at the sudden appearance of single circles or concentric circles, which look somewhat like eyes, but they were much less affected by the appearance of crosses or pairs of parallel lines. The greatest startling effect was obtained by projecting two realistically eyelike patterns that looked remarkably like the eyes of a cross-eyed owl.

Some of cecropia's close relatives, such as the polyphemus moth, have well developed eyespots on their wings—especially on the hind wings. Cecropia has large but only rudimentary eyespots on its front and hind wings, but it can probably make an effective startle display. Cecropia rests with its wings held together up over its back. If it is touched with a pencil point, it spreads its wings wide to reveal the colorful red and creamy-white upper surfaces with their rudimentary but boldly visible eyespots.

While birds seem to perceive large eyespots as a threat from a vertebrate predator, they seem to perceive very small eyespots as targets. When they peck at insects, they aim for the head end rather than the tail end. That way the insect is less likely to escape. Looking for the eyes is, of course, a good way to tell the head end from the tail end. Blest did an experiment to determine the effect of small eyespots, presenting yellow buntings with mealworms that had been painted with small eyespots on their tail ends. Birds presented with unpainted mealworms directed most of their pecks at the true head end. Birds given mealworms with an eyespot on the tail end directed most of their pecks at the tail end.

Some insects have inconspicuous true heads but have conspicuous "false heads" on their tail ends. A lanternfly from Thailand, actually a true bug with a sucking beak, rests with its head bowed low, a position that hides the beak, the antennae, and most of the rest of its head. On its tail end, which is somewhat elevated when the bug is at rest, is a remarkably convincing false head. There are two eyespots and a "beak" and a pair of "antennae" that are actually appendages of the tips of the wings.

Many of the hairstreak butterflies, including several North American species, have false heads on their hind wings. Eberhard Curio, who was once Konrad Lorenz's student, made some close observations of an Ecuadorian hairstreak. The tips of the hind wings of this butterfly have thin tails that look like antennae, and just behind them are small eyespots. Instantly after it lands from a flight, this butterfly swiftly flips around so that its tail end is where its head would have been. Then, as the real antennae are held motionless, slight movements of its hind wings cause the false antennae to wiggle.

Although there is little evidence to support it, a plausible hypothesis has been advanced to explain the occurrence of false heads on so many insects. The argument is that the false head deflects the peck of a bird away from the more vulnerable true head. A bird that is thus deceived makes only a superficial wound and comes away with nothing more than a bit of the wing rather than with the whole insect. The insect further confuses the bird by jumping or flying off in the direction opposite to what the bird expects. As Wolfgang Wickler wrote in his book *Mimicry in Plants and Animals,* "Two heads are better than one."

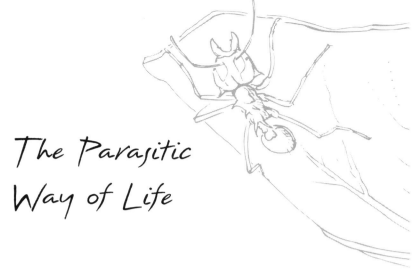

The Parasitic Way of Life

My teacher brought a cecropia cocoon to school when I was in the third grade and eight years old. She kept it in a cold place near a window, in a glass jar with holes punched in the lid. She told us what it was and that a large red and creamy-white moth would eventually emerge from it. The moth appeared in the early spring, but it was more beautiful and much larger than I had expected it to be. Its wing span was over five inches. It was the biggest insect I had ever seen.

In the winter of 1940, when I was twelve years old, I found a cecropia cocoon of my own. I put it and the apple twig to which it was attached in a glass jar with holes in the lid and kept it near a window at home. Nothing emerged from it until the following summer, and what did emerge was definitely not a moth. It was a large wasp with a long tail almost as thin as a horse hair. This unexpected outcome puzzled me and I went to consult my friend, Joe Brauner, a naturalist who worked part time at the Barnum Museum in my hometown of Bridgeport.

Since you have probably already guessed what Joe told me, I will leave that until a bit later and first tell you something about the fascinating history of the Barnum Museum. P. T. Barnum was probably the greatest American showman of all time. He is best known for his three-ring circus, but he also operated a commercial museum in New York City. Barnum lived in nearby Bridgeport, Connecticut, in a mansion called Iranistan that he had built near the shore of Long Island Sound. During the warmer months, he kept a man with an elephant endlessly plowing a field near Iranistan to advertise his show to passengers on the New York, New Haven, and Hartford trains that passed nearby. The mansion burned down long ago, but to this day there is an Iranistan Avenue in Bridgeport. After his New York museum burned in 1868, Barnum donated to his hometown a large stone building, especially erected to house the remnants of his collections.

His New York museum had had live exhibits, including Chang and Eng, the original Siamese twins, and Charles S. Stratton, also known as General Tom Thumb—only twenty-five inches tall and born in Bridgeport on what is now called Tom Thumb Street. There were no living exhibits in the Bridgeport museum, and, by the time I knew it, the Board of Education had taken over the lower floors of the building, crowding the museum into the small third floor and a storage loft. In more recent years, the city restored the museum, returning to it all of the original building and even adding on a new wing. Nevertheless, even the curtailed museum of the 1940s was heaven to a budding naturalist like me. In addition to the inevitable mementos of Tom Thumb, there were natural history exhibits, Indian relics, and rocks. Best of all there were Miss Clara Osborn, the curator, and Joe Brauner, both always willing to talk to a small boy and let him do odd jobs around the museum.

To get back to my cocoon, Joe told me that the insect that had emerged from it was an ichneumonid wasp, an internal parasite of cecropia caterpillars. This was the first time that I had ever heard of insects that parasitize other insects. It was an amazing revelation.

When my parasitized cecropia was a young caterpillar, a female ichneumonid like the one that emerged from my cocoon used her ovipositor, which in her species looks like a long, hairlike "tail," to inject one of her eggs into its body. She probably laid other eggs in other cecropia caterpillars and perhaps in caterpillars of polyphemus, promethea, or other moths closely allied to cecropia. At first the ichneumonid larva did little or no injury to its host, growing slowly in the body cavity of the caterpillar, absorbing nutriment from its blood and nibbling at its fat cells. It did

no serious injury until after the caterpillar finished growing and had spun its cocoon. Then the parasite devoured the entire caterpillar and spun a thick-walled cocoon of its own within the cecropia cocoon. It spent the winter as a pupa in diapause and did not emerge until well into the following summer, when partly grown cecropia, promethea, and polyphemus caterpillars were again available to be hosts for its offspring. If it had emerged earlier, at the same time that cecropia moths emerge, the caterpillars would have been eggs or still too small to serve as hosts.

Insects are commonly parasitized by nematode worms, mites, or one or more of the many insect species that live as parasites in or on other insects. There are also many microscopic viruses, bacteria, protozoa, and even fungi that inhabit insects. We often think of these microorganisms as agents of disease that should be put in a different category than the much larger parasitic worms, mites, or insects. But they do fit the definition of a parasite; they take their nutriment from the host's body without necessarily killing it, and they are, of course, smaller than their hosts. Furthermore, the larger parasites—nematodes, mites, and insects—cause disease just as do tiny microorganisms that are invisible to the naked eye. For example, the bites of body lice of humans can transmit the germ that causes typhus, but the bites of the lice in themselves cause injury and discomfort—in other words, disease, a word that comes from the Middle French *desaise:* literally, the opposite of *ease.* The disease caused by the lice themselves is known as pediculosis—from the Latin *pediculus,* a louse.

While a few of the insects that live in or on other insects are parasites in the true sense of the word—like the nematode roundworms that absorb nutriment from a pig's intestine but do not usually kill it—entomologists point out that the great majority of parasitic insects, including the ichneumonid wasp from the cecropia cocoon, should really be called *parasitoids,* because they behave both as parasites and as predators. In its early stages, an ichneumonid larva, for example, takes nutriment from its host without seriously injuring it—behavior typical of a parasite. But after the host spins its cocoon, the ichneumonid larva devours and kills it—behavior typical of a predator. To keep things simple, I will just use parasite as a general term to refer to parasitoids and true parasites taken together.

Many insects are attacked by more than one kind of parasite. Cecropia caterpillars, for example, may be attacked by the parasitic larvae of at least nine different species of insects, as well as by various microorganisms. The insects total four species of ichneumonids, including the one that emerged from my cocoon; there are two other parasitic wasps; and there are three different kinds of tachinid flies. The parasitic insects that infest cecropia caterpillars are in turn attacked by parasites of their own, and there may even be parasites of the parasites of the parasites. Jonathan Swift, the eighteenth-century British satirist, wrote:

> So, naturalists observe, a flea
> Hath smaller fleas that on him prey;
> And these have smaller still to bite 'em;
> And so proceed *ad infinitum*.

As the next two lines make clear, Swift was actually voicing a complaint about literary critics:

> Thus every poet, in his kind,
> Is bit by him that comes behind.

Nevertheless, his quatrain expresses the biological idea succinctly and with wit.

Fleas are not really infested by smaller fleas, but as in the case of cecropia, the parasites of insects are often attacked by yet smaller parasites that biologists refer to as hyperparasites. The chain of hyperparasites upon parasites upon host is not, of course, infinitely long as Swift says, but there are cases in which the chain may have as many as five links. In 1937, Frank L. Marsh published a report of his master's degree research at Northwestern University. He had discovered that the caterpillars in a large cecropia population in southern Chicago were attacked by two of the nine parasites that I mentioned above: an ichneumonid wasp and a tachinid fly. These two parasites were attacked by several hyperparasites. The tachinid chain includes four links: cecropia; the

tachinid that parasitizes it; a wasp parasite of the tachinid; and another wasp that parasitizes the parasite of the tachinid. The ichneumonid chain, however, has five links: cecropia; the ichneumonid wasp that parasitizes it; a second ichneumonid species that is a parasite of the first one; a chalcid wasp that parasitizes the second ichneumonid; and a second chalcid species, which parasitizes the first chalcid.

The *Coccophagus* wasps are minute parasites, some of which may be less than four one-hundredths of an inch in length. The literal translation of their name, given to them before their habits were fully understood, is "eater of scale insects." We now know that this name is not totally accurate. The females do indeed parasitize tiny scale insects, but the males of some species have different hosts than do their sisters. They may be hyperparasites of parasites that attack scale insects or they may even be parasites of the eggs of moths. Sometimes the males are hyperparasites of members of their own species.

Coccophagus mothers place their eggs in an appropriate host, in the body of a scale insect if the egg is destined to become a female or in the egg of a moth or the body of a parasite of a scale insect if it is destined to become a male. Like the female honey bee, the wasp can choose to fertilize or not to fertilize each egg that she lays. Fertilized eggs become females and unfertilized ones become males. Thus the ovipositing mother knows the sex of each egg that she lays. Unfertilized females lay only male eggs and place them accordingly. Fertilized females may lay either male or female eggs, but usually lay only fertilized female eggs and place them in the body of a scale insect.

Parasitizing other insects is a way of life for many insects—for at least 115,000 species. It is their ecological niche, the way in which they obtain food and make a living. About 100,000 of these parasites are various species of wasps, and about 11,000 of them are flies—the majority of them tachinid flies. Most of these insects are parasitic only in the larval stage. The adults are usually free-living and feed on nectar and honeydew, although a few of them lap the juices of the host insect after they have pierced it with their ovipositors. Many parasites live within the body of the host, but some cling to the outside of the body and use their mouthparts to pierce the body wall of the host so as to obtain nutriment.

If there is an insect species that does not have parasites of one sort or another, I have yet to hear about it. Although no species is completely immune, at least a few members of all species either escape or survive the attacks of parasites. Insects have evolved numerous ways of avoiding or discouraging parasites, ranging from aggressive defenses to shifting to new habitats that have not yet been occupied by the parasites that attack them.

Some insects may struggle violently in order to fend off adult parasites that are about to lay an egg in or on their body. Some aphids kick with their long legs, while others escape by allowing themselves to fall to the ground when a parasite is about to lay an egg in them. Many caterpillars writhe and thrash about violently and thus sometimes dislodge a parasite before it can lay an egg, and other caterpillars take advantage of their flexible bodies in order to strike back and snap at adult parasites that land on them. The sharp spines or long hairs that cover the bodies of some insects are obstacles that make it difficult for a wasp to reach down to the skin and pierce it in order to inject an egg. The same purpose is served by the waxy filaments that cover the bodies of some aphids. The armor plate of hard-shelled beetles is difficult for parasites to pierce.

Although larval Mexican bean beetles are soft-bodied, they are covered with thin, multibranched spines that are known to protect them from predaceous ants and that probably protect them from parasites too. When an insect brushes against one of these larvae, the fragile spines break and leak a sticky fluid that gums up the intruder's body. The intruder limps away and attempts, often in vain, to clean its body.

The pupae of some beetles and those of a few moths are protected by clamplike devices called gin traps. These pincerlike traps are located over the flexible membranous areas between the hard plates of the abdominal segments. When the abdomen is at rest the trap is open. When the pupa writhes its abdomen, the

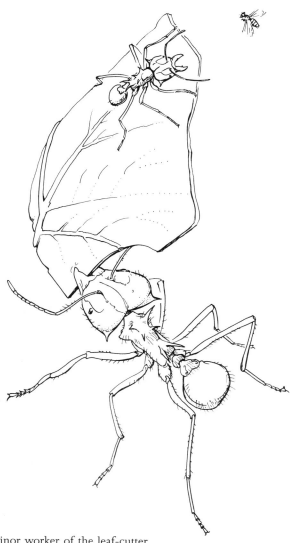

A minor worker of the leaf-cutter
ant defends the larger, leaf-carrying
worker from an attack by a parasitic
phorid fly

abdominal segments move against each other and the two jaws of the trap alternately close and open and may pinch the leg or even the ovipositor of an intruding parasite.

Leaf-cutter ants of the New World tropics use a remarkable cooperative method to defend themselves against parasitic phorid flies. Long columns of leaf-cutter workers carry pieces of leaf back to the nest, where they will be chewed into a mulch on which these agricultural ants grow a fungus that is their staple food. Each worker carries a piece of leaf that is many times its own size, holding it aloft like a huge green sail. The parasitic flies hover over the column, looking for an opportunity to dart in and lay an egg on the neck of one of the workers. After hatching, the phorid larva feeds in the ant's head and ultimately kills it by eating the brain. The leaf-carrying worker ants are so laden that they cannot defend themselves. However, as detailed by Irenäus and Eleonore Eibl-Eibesfeldt, they are often accompanied by a guard, a very small type of worker that is about one-third of their own length. The tiny guard sits on the leaf-piece, riding shotgun. She snaps at intruding flies and often succeeds in driving them off.

Some insects build shelters that tend to protect them against the attacks of both parasites and predators. The large silken nests of the eastern tent caterpillar of North America are often seen on wild black cherry trees growing along country roadsides. In early spring the caterpillars, which overwinter as masses of eggs on the twigs of the tree, move to one of the main forks of a branch and there spin a sturdy and tightly woven silken communal nest. They spend the night in the nest but leave it to feed on foliage during the warm daylight hours. As the tent caterpillars grow, they enlarge their nest until it may be as much as two feet long. Although some parasites and predators manage to enter these nests, many of them are baffled by the silken walls. Thus the caterpillars are at least partly protected during the cool parts of the day, late afternoon, night, and early morning.

The fall webworm, an insect common throughout most of the United States and southern Canada, builds an even larger communal nest that is sometimes mistaken for that of the eastern tent caterpillar. The webworm nest, however, is not built in the fork of

a main branch. It is, rather, built at the tip of a branch and, unlike the nests of tent caterpillars, it encloses the foliage that its inhabitants eat, and is enlarged to include more and more foliage as the season progresses and the caterpillars' size and appetite increase. Since they do not have to leave the nest, fall webworm caterpillars may be better protected from parasites and predators than are tent caterpillars.

Other insects, including a few caterpillars and many aphids and their relatives, are protected from parasites by the ants with which they associate. This is usually a mutualistic relationship from which both the guarded insects and the guarding ants derive benefit. As you have already seen, aphids reward the ants that guard them with honeydew, their sugar-rich excrement. Caterpillars do not excrete honeydew. But many of those that are guarded by ants have "honey glands" that secrete a sweet and nutritious exudate that the ants gobble up as greedily as they do the honeydew of aphids.

Honey glands are borne on the upper side of the abdomen of certain members of the Lycaenidae, the gossamer-winged butterfly family, the familiar and dainty, little "blues" and the hairstreaks you have already encountered. This gland, sometimes called Newcomer's gland, was first described in the entomological literature by E. J. Newcomer in 1912. At that time it was only assumed that the secretion of this gland is nutritious. It was not until 1975 that U. Maschwitz and two colleagues published a chemical analysis of the secretion, showing that it consists of from 13 percent to almost 19 percent sugars and that Newcomer's gland is truly a "honey gland" that provides nutrients for the ants that imbibe from it.

In a 1981 issue of *Science*, Naomi E. Pierce and Paul S. Mead presented the first data that prove conclusively that caterpillars, specifically those of the silvery blues, lycaenid butterflies that are found throughout much of southern Canada and the United States, actually do benefit from being tended by ants. In a random survey they collected caterpillars that were or were not tended by ants. About 63 percent of the caterpillars that were not tended or that were tended only by an individual ant were parasitized by insects. But of those tended by three or more ants, less than half as many, only 26 percent, were parasitized. In an experiment, some caterpillars fed on plants from which ants were excluded by a band of Tanglefoot, a sticky substance used to trap insects, on the stem; others fed on plants that were ascended by ants because the band of Tanglefoot did not completely encircle the stem. About 36 percent of the larvae from plants patrolled by ants survived, but only about 17 percent of those not guarded by ants survived. Furthermore, when the surviving larvae were allowed to complete their development, 42 percent of those that had not been tended by ants were found to be parasitized, but only 18 percent of those that had been tended by ants were parasitized.

In their Pulitzer Prize–winning book of 1990, *The Ants*, Bert Hölldobler and Edward O. Wilson pointed out that the egg-laying mothers of ant-associated caterpillars are much more likely to land on and to lay their eggs on plants that are frequented by ants than on plants that are free of ants. They went on to say that the more dependent a lycaenid species is on its attendant ants, the more likely it is to lay its eggs on plants that are occupied by ants.

Some insect populations escape parasites by shifting frequently to new ephemeral habitats, thus keeping a step ahead of the parasites. The idea is to arrive in a new habitat before the parasites do and to shift to another new area after the parasites arrive. As I said earlier, insects that practice this form of escape, cecropias among them, are known as fugitive species because their populations are constantly on the run to escape predators, parasites, or disease-causing microbes.

Just what are ephemeral habitats, and how do they come to be? The life spans of habitats vary, but there are many that can be said to be ephemeral, to persist for only a short while. A cow pat is a suitable habitat for maggots only when it is fresh, for some species for only a few days. Temporary woodland pools formed by the spring rains dry up in the heat of summer. But during the few weeks that they persist, they are ideal habitats for a host of aquatic insects and crustaceans, including the odd little fairy shrimps. The ecological succession that restores an area after a farm field is abandoned or after a woodland is denuded by a forest fire or a landslide consists of a series of ephemeral habitats that may endure for a year or two or for a decade or more. A denuded former woodland is first colonized by various herbs and grasses; after a few years they are succeeded by mixed herbs and shrubs, which are in turn replaced by short-lived trees such as hawthorns and cherries. The last constitute the preferred habitat of cecropia. These short-lived trees are ultimately succeeded by oaks, maples, and other long-lived trees, the climax species that will continue to replace themselves and thereby persist until the land is next swept by fire, landslide, or the axe and the plow.

In 1981, Jim Sternburg, Aubrey Scarbrough, and I published research results which demonstrate that cecropia is a fugitive species. Aubrey did most of the field work, and has since gone on to become a professor at Towson State University in Baltimore. Our data show that cecropia is almost absent from mature, climax woodlands in central Illinois, but that it occurs more commonly in areas that have characteristics of an early stage in a woodland succession: roadsides with stands of willows and wild cherries and recently built suburban housing developments with sapling shade trees such as birches, hawthorns, and silver maples.

How do cecropia populations make the shift from one ephemeral habitat to another? The female probably does not have the ability to assess the abundance of parasites and, if there are too many of them, postpone laying eggs until after she has flown off and located another ephemeral habitat that is not yet occupied by parasites. But her instinctive behavior sometimes accomplishes essentially the same end. After laying some eggs in the area in which she was born, she flies off and lays the rest of them elsewhere. The female is heavily laden with eggs, and how far she can fly is not known, but if she is anywhere near as capable as a male, she could fly several miles in a night and move a distance of many miles even during her short lifetime.

By good fortune, she may arrive at another ephemeral habitat, one occupied by short-lived trees, that suits her requirements and that, if luck is with her, has not yet been found by parasites. There she will lay some of her eggs, although she does not know whether or not parasites are present. The larvae that hatch from the eggs that she laid in her original habitat may well be doomed to become food for parasites, but her children and perhaps several succeeding generations may flourish in the new habitat until the parasites catch up with them. Not every wandering female will locate a suitable new habitat that is free of parasites. When most of the eastern United States and southern Canada were covered by forests, early stages of the ecological succession were scattered and few, because there were relatively few disturbed and denuded areas. Today such areas or similar ones are more common because of human disturbance, and it may be that cecropia is consequently more abundant than it has ever been.

Insects have one last line of defense after a parasite has penetrated their bodies. Many species can encase the egg or young larva of a parasite in a capsule that renders it helpless and ultimately smothers it to death by blocking its access to oxygen. The capsule is formed of blood cells that migrate to the parasite and flatten themselves closely against its body. They eventually gather in such numbers that they form a many-layered, closely fitting capsule that completely surrounds the parasite. Insects can similarly encapsulate almost any foreign object that is inserted into their bodies. There is one major exception. They usually cannot encapsulate a parasite that is specialized to attack them and other members of their species.

Many parasites are highly specific in their choice of hosts. Some attack many kinds of insects, but others attack only insects of one species, and many attack only a few closely related species. These specialist parasites have evolved ways of preventing their usual hosts from encapsulating them. But they generally have not evolved the ability to prevent all potential hosts from doing so. Thus host insects can often encapsulate a parasite that is not specialized to cope with their defenses, but are often unable to encapsulate a parasite that is specialized to deal with them.

Simple experiments, done by George Salt of Cambridge University in England, showed that the mother parasite of specialist species coats her eggs with something that prevents encapsulation by the usual host. When eggs that had been laid in the normal manner were artificially implanted in hosts, they were not encap-

sulated. But eggs that were removed directly from the ovary and implanted in a host were encapsulated. These latter eggs had not passed down the oviduct, the egg tube into which the accessory glands of the female genitalia release their secretions. The obvious conclusion is that eggs that are laid normally and pass down the oviduct are coated with a protective substance or substances secreted by the accessory glands.

This conclusion is supported by the results of a different experiment that proves that there is a coating on normally laid eggs by showing that it can be removed. Some eggs that had been laid in the normal manner were abraded or washed with solvents by the experimenter. These eggs were not protected against encapsulation. Other eggs that had been laid in the normal manner were implanted in the host without any attempt to remove the protective coating. These eggs were protected against encapsulation.

Only a few parasitic larvae can locate hosts on their own, and it is usually the parasite's mother that seeks out the host and places her offspring in or on it in the form of an egg. Not just any insect will do. If the parasite is to survive, the host must belong to the right species, and it must be of the right age, usually an immature form. Eggs, larvae, pupae, or nymphs are often parasitized, but adults are relatively seldom parasitized. As you have already seen, the great majority of parasites are host-specific to varying degrees; that is, their evolutionary adaptations specialize them to parasitize only one species, a small group of closely related species or sometimes a larger group of more distantly related species. No parasite is capable of living in or on all or even most insects.

How, then, does a female parasite find appropriate hosts for her offspring from among the many different insects that she might encounter? Only a few parasites have been closely studied with regard to this question. Much more research is needed to provide a comprehensive answer, but enough is now known to suggest a few tentative generalizations. First, host finding is generally governed by a combination of factors that probably form a hierarchy of clues—with each succeeding clue bringing the female closer to the host and to the decision to lay an egg.

Some parasites are known to seek out the host's habitat as a first step in locating the host itself. When a female parasite first emerges as an adult, she is, of course, in the habitat of her host. There may, however, be good reason to abandon her natal habitat and to search for another similar one in which to lay her eggs. In other words, it may be beneficial to be a fugitive species, like cecropia. Her original habitat is likely to be occupied by many other members of her own species. Thus, her progeny will avoid competition and fare better if she disperses to a new habitat not yet occupied by members of her own species and lays her eggs there. The odor of the plant on which the host insect feeds is often the clue that ovipositing parasites use to find the right habitat. The smell of an uninjured plant can suffice, but as we will see later, the odors released by a plant on which the host has been feeding may play a special role. As a case in point, newly emerged females of an ichneumonid that parasitizes caterpillars of the European pine shoot moth are repelled by the odor of pine oil. They are not yet mature enough to produce eggs. About three weeks later, their ovaries mature and they are ready to lay eggs. Then they are attracted to the odor of pine oil, and as they move about they may be lucky enough to chance on a stand of pines infested by shoot moths that are not yet being exploited by parasites.

Once a gravid female parasite arrives in the proper habitat, she will most likely be guided to host insects by her sense of smell. The guiding clue may be the odor of the host itself, often one or more of its pheromones. It may be the odor of the host's fecal material or sometimes even the odor of silk spun by the host. A parasite of the infamous gypsy moth is attracted to the odor of the silk spun by its hosts but not to the silk of other caterpillars. The odor emanating from plants damaged by herbivorous hosts is often attractive to parasites. When a parasite locates a potential host, tactile, gustatory, and close-range olfactory clues will confirm its identity. Not until then does the mother parasite lay one or more eggs in or on the host.

A few parasitic insects may use visual clues to locate their hosts from a distance, but there is not much evidence to prove this point. The clue is probably seldom the sight of the host itself, since the limited visual capacity of insects may not be enough to resolve something as small as another insect in sufficient detail. The parasites are more likely to clue in on something more visible, such as the extensive damage that their hosts do to the leaves on which they feed. A leaf-feeding caterpillar or beetle larva, for example, may damage several leaves, eating numerous holes into each one and removing well over 50 percent of the area of each leaf. Some caterpillars eliminate such obvious signs of their presence by clipping the stems of the leaves on which they have fed and thus letting them fall to the ground. M. P. Hassell of the University of Oxford presented evidence which suggests that a tachinid parasite uses damaged leaves as a clue to the presence of its hosts, the caterpillars of the winter moth. The more damage that was visible on an oak, hawthorn, or hazel, the greater the percentage of caterpillars on the tree that was parasitized by maggots of the fly. The parasite, however, might actually have been responding to the odor of the damaged leaves, rather than the sight of them.

Even as I write, entomologists are learning more and more about an unforeseen chemical route by which parasites are led to their hosts. Accumulating evidence demonstrates that at least some plants send forth a cry for help when they are under attack by plant-feeding insects. When injured by insects, these plants release scents that attract and bring to the rescue parasites and predators of their attackers. As long ago as 1955, the Canadian entomologist L. G. Monteith had shown that parasitic insects are attracted to plants that have been damaged by their hosts. But it was not until 1990 that T. C. J. Turlings and two of his colleagues, all research entomologists working with the U.S. Department of Agriculture, discovered a new and amazing aspect of this line of communication between plants and insects. In an article in *Science*, they showed that a plant does not send out a call for parasites unless it is actually under attack by plant-feeding insects.

Turlings and his colleagues found that plants that have been chewed on by armyworm caterpillars release volatile chemicals, mainly terpenoids, that attract certain parasites of the caterpillars. The plants release these attractants only if they bear wounds that have been contaminated with oral secretions of the caterpillars. Wounding alone is not enough. Plants that were artificially damaged with a razor blade did not release the terpenoids and did not attract parasites. But when the wounds on artificially damaged plants were smeared with regurgitant from armyworms, the plants did release the volatile chemicals and did attract parasites. Uninjured plants smeared with caterpillar regurgitant did not release the attractive volatiles and did not attract parasites. Clearly, both wounding and contamination with oral secretions, as occur when an insect chews on a plant, are required to trigger the release of the chemicals that attract the parasites. Thus the plants communicate a specific message to the parasites: Come to me and you will find hosts for your offspring.

As R. R. Askew points out in his authoritative book, *Parasitic Insects*, the mothers of most parasites lay their eggs in or on the bodies of the hosts that their larval offspring will occupy. Their newly hatched progeny are maggot-like, legless, and incapable of moving from place to place; they could not possibly locate a host on their own. On the other hand, some parasite mothers lay their eggs some distance from a host. Their newly hatched larvae do have legs or some other means of locomotion and can and must find a host by themselves.

Certain relatives of the blow flies, notably the cluster fly, are unusual among the insects in that they are internal parasites of earthworms. The females lay their eggs in crevices in the soil, and the newly hatched maggots, which are quite mobile, search out the earthworms. Perilampid wasps are also among the parasites that do not lay their eggs in or on the host. Among them is a species that, in its larval stage, is an external parasite of larval and pupal green lacewings. The larvae of green lacewings, known as aphidlions, are voracious predators that move through aphid colonies sucking the body fluids from one aphid after another and tossing away the dry shells. The wasps increase the likelihood that

their offspring will find an aphidlion host by sticking their eggs to foliage infested by aphids. The mobile and long-lived newly hatched larvae are legless but move about on the leaf like inch worms, and eventually fasten themselves to the leaf surface by means of a fleshy sucker at the terminal end of the abdomen. The attached larva rears up at a right angle from the leaf surface and waits for a host to come near. At the approach of any moving thing, signaled by vibrations of the leaf, the tiny wasp larva reaches out and twists about in a frantic effort to make contact with it. Upon contact, the parasite grabs on firmly with its mandibles. If it finds itself on an aphidlion, it will hold on until the aphidlion pupates. Then it will begin to feed on the pupa and will eventually kill it. If it attaches to some creature other than an aphidlion, it will ultimately die.

Most of the other perilampids are hyperparasites of tachinids, ichneumonids, or other insects that attack caterpillars. Like the perilampid parasite of aphidlions, these species also lay their eggs on foliage rather than in or on the host, and their newly hatched larvae must do the job of finding the host. One of these species is a hyperparasite of another wasp, a braconid that is an internal parasite of caterpillars of the grape leaf folder. The mother parasites lay their eggs on grape foliage. The newly hatched larva attaches itself to a passing caterpillar and bores into its body and searches for a braconid parasite. If no braconid is present, the perilampid larva usually dies. If a braconid parasite is present, the hyperparasitic larva burrows into its body and remains there until the braconid makes its way out of the caterpillar's body to spin a cocoon and pupate. Then the hyperparasite emerges from the braconid's body and finishes its development as an external hyperparasite of the braconid pupa.

Some parasitic insects are phoretic. Phoresy, using an animal of another species to provide transportation, has been called "biological hitchhiking," but it is really more like a hobo's hopping a freight train. Unlike a hitchhiker, the phoretic creature does not ask permission to come on board and is certainly not invited to take a free ride. Wasps of the family Eucharitidae, parasites of ant larvae and pupae, are a case in point. The females stick their eggs in shallow slits that they cut in the leaves of plants, usually those of plants that are infested by aphids. The newly hatched larvae attach themselves to worker ants that are attracted by the aphid honeydew. The parasitic larvae are carried back to the nest by the ants. There, the parasite attaches itself to an ant larva, but it does little or no feeding on it until the larva is fully grown and metamorphoses to the pupal stage.

Some insects that are parasites of insect eggs or larvae are phoretic as adults. Larval scelionid wasps, for example, parasitize the eggs of insects. The winged female of a species that attacks the eggs of mantises attaches itself to a mantis, usually a female, and clings to the abdomen or the wing base of the mantis with her mandibles. Soon after boarding the mantis, she discards her wings and thus commits herself for life. Not only does she hitch a ride on the mantis, but she also sucks its body fluids as she waits for the mantis to lay its eggs. She is one of the few adult insects that is a parasite of another insect. If the mantis dies, she dies with it. The mantis will probably eventually lay a large clutch of eggs and cover them with a fluid, frothy secretion that hardens to form a sturdy, protective case. When the mantis begins to lay its eggs, the parasite quickly moves to them and inserts her eggs into them before she is prevented from doing so by the hardening of the secretion.

The mantispids are close relatives of antlions and lacewings. But adult mantispids look more like praying mantises than like their own relatives. The striking similarity between these two groups came about because both groups—quite independently of each other—evolved almost identical ways of solving the same problem, the capturing of prey. Biologists call this convergent evolution. Both mantispids and mantises are ambushers rather than stalkers or pursuers of prey. They sit quietly on a plant and wait for prey insects to appear. When an insect comes close enough, they reach out to seize it with front legs modified for grabbing. Their success rate is increased by their extended reach, which results from the great lengthening of the thoracic segment that bears the front legs and of the legs themselves. Both mantispids

This mantispid, though not related to the praying mantis, has independently evolved a method of capturing prey, in this case a house fly, that is similar to the mantis's

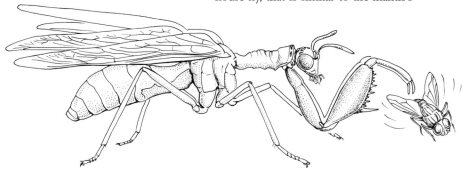

and mantises look rather like giraffes, with a pair of long grabbing legs growing out of the neck from just behind the head.

Although adult mantispids are predators of other insects, as larvae they live and feed within the silken egg sac of a spider. Once every few days, an adult female mantispid lays a clutch of from several hundred to several thousand eggs, each attached by a short stalk to some surface in the environment. The number of eggs laid by a female depends upon her size, which in turn depends upon the amount of egg material that was available in the spider egg sac in which she grew to maturity. The newly hatched mantispid larva has legs and moves about quite actively. It will climb on board almost any spider that it encounters, gripping the spider tightly and sucking its blood. If the spider is a female, the mantispid larva crawls into her egg sac as she spins it. But all is not lost if the spider is a male. Larvae on male spiders may transfer to a female that cannibalizes her mate after copulation, or if the male spider dies for some other reason, the mantispid larva will leave its body and may be able to board another spider that may turn out to be a female. When the mantispid larva matures, it spins its own silken cocoon within the egg sac, pupates and, after it emerges as an adult, chews its way out of its cocoon and the egg sac.

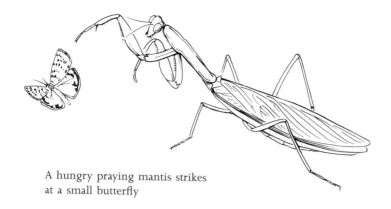

A hungry praying mantis strikes
at a small butterfly

The sloth moths of the New World tropics provide one of the most unusual examples of phoresy. The adult moths live in the hair of both the three-toed and the two-toed sloths. These slow-moving mammals have been found to be occupied by anywhere from 19 to 132 moths of both sexes. The moths were discovered in the nineteenth century, but until 1976 not much was known about them. It was assumed that eggs, caterpillars, pupae, and adult moths all live in the hair of sloths. Several authors speculated that the caterpillars feed on the algae that grow on the hair of the sloths or that they eat the hair itself.

However, in a 1976 article in *Science*, Jeffrey K. Waage and G. Gene Montgomery reported that, although they found many adult moths on sloths, they found no eggs, caterpillars, or cocoons of these moths on fifteen sloths that they examined. They did, however, find caterpillars feeding on the dung of sloths. Sloths eat the leaves of trees high in the forest canopy, but about once a week they descend to the forest floor to defecate. They hang from a vine by their front legs and use the claws of their hind legs to dig a hole in the soil. Then they deposit about a cupful of fecal pellets in the hole and cover them with leaf litter. At this time, female moths leave the sloth and lay their eggs on its dung. The larvae feed on the dung, and when the adults emerge from their pupae in the dung pile, they fly up into the canopy to seek out a sloth, in whose fur mating occurs. Gravid females leave the sloth only when it has defecated on the ground.

⬎ Although most parasitic insects attack insects, some of them parasitize other animals, ranging from sponges to human beings. We heard about some of them in preceding chapters: the human bot fly or tórsalo; the fleas and lice that attack humans, other mammals, and birds; and the sheep ked that spends all of its life stages in the wool of a sheep or the hair of a goat.

Before we go on, let us return to the subject of the sheep ked. In their recent text, *Destructive and Useful Insects*, Robert L. Metcalf and his son Robert A. Metcalf describe the unusual ways of this insect. So far I have said very little about this aberrant insect and its unusual life style. Although sheep keds are definitely not arachnids, they are often called sheep "ticks" because of their resemblance to the true ticks. Like ticks, they drink the blood of their hosts, they are always wingless, their bodies are flattened, and their abdomens are unsegmented and sac-like. Sheep keds are actually highly specialized flies that have lost their wings and have otherwise become adapted to spend their entire lives on the body of their host. They originated in Asia, presumably on the wild ancestors of domestic sheep, but are now found throughout the world wherever sheep are grown.

The overall mobility of the sheep ked is obviously diminished by the loss of its wings, but its mobility on a sheep is clearly enhanced. If it is to move from place to place on the sheep—as it must do—it has to make its way through the dense and tangled jungle of fleece that covers the sheep's body. Wings would hinder such movement, and might soon become so broken and frayed that they would be useless for flight. But the loss of the wings also has a serious disadvantage. How would a wingless adult sheep ked find and board a sheep if it spent the maggot (larval) stage somewhere other than on the sheep—perhaps in dung or carrion, as do most other maggots and as did the maggots of the ancestors of the sheep ked. Boarding a sheep would be a difficult climb for a wingless adult, and, anyway, the flock of sheep would probably have moved on by the time a free-living maggot could finish growing and emerge as an adult and crawl around looking for a host.

Sheep keds have solved this problem by not having free-living larvae. Like tsetses, they retain their larvae within their bodies, nourishing them with milk secreted in the "uterus," an enlarged area of the common oviduct, and giving birth to them when they are full grown and ready to become pupae. During her lifetime, a female sheep ked may give birth to ten or twenty maggots that are nurtured in her uterus one at a time. Each newly born larva is firmly glued to the hairs of the host by the mother.

The helpless and immobile pupa can remain on the sheep because of the unique way in which it and the maggots of all the higher flies, as contrasted with the more primitive flies such as mosquitoes and gnats, prepare for pupation. Before it pupates, the carrot-shaped maggot assumes an oval shape and darkens and hardens its soft white skin. Only then does the pupa separate itself from its larval skin. But it does not shed its larval skin. It remains within it, protected by it as a cecropia pupa is protected by its cocoon. The pupa is thus retained on the sheep's body by the glued-down larval skin. When the adult fly is ready to emerge from the puparium, as this hardened larval skin is known, it bursts its way out with a large inflatable bladder that protrudes through an opening at the front of the head. After emergence, the bladder is withdrawn into the head and the opening closes.

Ox warbles do not make music. Their name is akin to the now obsolete Swedish word *varbulde*, a boil—a compound word that is formed from *var*, pus, and *bulde*, a swelling. The ox warbles, also known as cattle grubs, are unpleasant internal parasites of cattle. There are two species in North America, the bomb fly and the heel fly. Adult females of both species lay their eggs on the lower belly or the legs of a cow. The newly hatched larvae penetrate the skin and then spend five or six months wandering through the muscles and other tissues of the cow's body until they reach the back. There they form a large, puss-filled, open boil from which they eventually fall to the ground to pupate in the soil. The two species differ in how they lay their eggs. The bomb fly female buzzes loudly around the cow and repeatedly darts in to glue an egg to a hair. The cow becomes terrified, raises its tail straight up, and dashes about wildly until it loses the fly by running through brush or entering the water. The heel fly is much more tactful. It does not frighten the cow. It lands in the shadows near the animal and then sneaks quietly over the ground to lay its eggs on the heels of a standing cow or on the belly of a recumbent cow.

Insects have even evolved to parasitize sponges—not saltwater species such as the bath sponge, but the few species that live in freshwater ponds, lakes, and streams. Among the insects that have evolved to exploit sponges are a few true bugs, fly maggots, caddisfly larvae, and larvae of the spongeflies. The latter are not really flies at all; they are relatives of the green lacewings and the mantispids. Spongeflies lay their eggs on leaves overhanging the water or on snags protruding above its surface. The newly hatched larvae drop to the water and drift about below the surface. Their gills remove dissolved oxygen from the water. Most of them fall by the wayside or are eaten by some aquatic predator, but a few lucky ones come into contact with a sponge and feed on it as they cling to its outer surface or even the inner surface of its body cavity. When they are fully grown, the larvae leave the water and spin a cocoon on a nearby plant or some other object.

A few animals, including some insects, are kleptoparasites: they steal food from other animals. This form of parasitism has also been called pilfering or piracy. Kleptoparasitism is generally rare, but it is fairly frequent among birds. Over 200 birds out of a world total of about 9,000 species are kleptoparisitic to varying degrees. Some even steal food from insects and spiders. Several hummingbirds, warblers, and a few others steal the prey from spiders' webs, and some of them, especially the hummingbirds, also take silk to use in the construction of their nests. American robins, Eurasian starlings, brown-headed cowbirds, and house sparrows steal prey that digger wasps carry to their burrows. S. S. Ristich described how robins and house sparrows stole grasshoppers from the digger wasps in a colony that had dug nests in a patch of soil near walkways adjacent to the Veterinary Building on the Cornell University campus in Ithaca, New York:

> At the Veterinary colony, the wasps were almost exterminated
> . . . by robins and English [house] sparrows who robbed the
> wasps of grasshoppers. The robins would patrol the walks until
> a wasp arrived, whereupon the robin would attack the wasp
> causing it to drop the grasshopper. The English sparrows not
> only attacked along the walk, but also used the Veterinary
> building as an observation perch. From this vantage point, they
> could see more wasps homing. Many sparrows, with the first
> grasshopper still dangling from their beaks, would chase other
> wasps for a second meal.

Only a few insects are known to be kleptoparasites. Among them is the death's head moth, a species of hawk moth whose name is derived from the skull-like pattern on its back. Occurring in Africa, Europe, and Asia, these moths drink nectar from plants but also rob honey from domestic hives or wild nests of honey bees. Like burglars, they steal into the hive at night and each one helps itself to about a teaspoonful of honey from uncapped cells. Dead moths that are sometimes found in commercial hives show that these thieves may be caught by the bees and given their just deserts. The bee louse, actually a tiny wingless fly that clings to the hairs on the body of a honey bee, is also a kleptoparasite. These tiny parasites look much like mites but have only six legs rather than the eight borne by mites. When they are hungry, bee lice move to the bee's proboscis and appropriate for themselves a share of the saliva and nectar regurgitated by the bee.

Karl Hölldobler described the behavior of a phorid fly that spends its life in ant nests and lives by stealing food from the workers. He reports that a hungry fly leaps onto the back of a worker ant, strokes the cheeks of the ant with its forelegs, and thereby stimulates it to regurgitate a drop of food. The fly licks up the drop and quickly slips away from the frightened and disturbed ant.

Harpagomyia, a mosquito with aberrant feeding habits, steals food from ants. In a letter communicated to a 1918 meeting of the Entomological Society of London by Professor E. B. Poulton, C. O. Farquharson of Ibadan, Nigeria, described how flying adult male and female *Harpagomyia* waylay ants:

> You know how worker ants stop each other and exchange a little regurgitated food, a momentary transaction almost, both passing quickly on their way. The mosquitoes do exactly the same. They will drop downwards just over an ant that is hastening along in the usual way. The ant may stop and give an alms to the beggar, passing on a moment or two later as if it had just met a friend, and the mosquito flies up and down again till another obliging ant is met. At times the selected ant simply ignores the mendicant, but shows no resentment, nor does the mosquito press his or her attentions.

As Randy Thornhill pointed out in a 1975 article in *Nature*, scorpionflies are scavengers that feed mainly on soft-bodied insects that are dead or moribund. They find much of their food by themselves, but they are also kleptoparasites that enter the webs of spiders to steal prey that are trapped there. The scorpionflies, close relatives of the hangingflies that I discussed earlier, either land on the web from flight or walk onto it from the vegetation to which it is anchored. They usually eat their booty as they sit on the web, but will fly off if they are approached by the spider. Although scorpionflies are agile and generally move about on the sticky web with little difficulty, they sometimes do become entangled. If they cannot otherwise extricate themselves, they regurgitate a brown liquid that dissolves the silken web. Scorpionflies are usually not caught by the owner of the web, but if they are, they regurgitate the same brown liquid and smear it on the spider. The spider then releases the scorpionfly and tries to clean itself of the deterrent liquid.

Certain bumble bees that belong to the genus *Psithyrus* are obligatory social parasites of other bumble bees. That is, they cannot survive unless they subjugate a colony of some other species of bumble bee. The parasitic species produce no worker caste of their own, only queens and males. The queens even lack the pollen baskets that nonparasitic bumble bees have on their hind legs. Since the parasitic queen cannot carry pollen, she could not start her own colony even if she wanted to. Consequently, these parasitic bumble bee queens cannot reproduce themselves unless they take over a colony of some other species of bumble bee, depose the rightful queen, and induce the workers to raise their progeny rather than those of the former queen. Some parasitic queens kill the host queen, but others force her to stop laying eggs and to assume the duties of a worker. If the host queen does manage to lay an occasional egg, the parasitic queen promptly destroys it.

Slavery is a form of parasitism. Although people who owned slaves probably did not think of themselves as parasites, they did forcibly appropriate for themselves the labor of other people. As Bert Hölldobler and Edward O. Wilson point out in *The Ants*, biologists have broadened the term slavery to include the exploitation of a species other than one's own. In the human sense, this is not true slavery at all. We think of the bees, sheep, horses, and camels that people exploit as domestic animals rather than as slaves. I will, nevertheless, use the word slavery in its broader, biological sense. Quite a few different kinds of ants are slave makers in the broad sense. They capture and exploit ants that belong to species other than their own. I will come back to them later.

But a very few ants—only two species are known so far—practice slavery in the strict sense. They capture and enslave ants of their own species, always individuals from foreign colonies, never members of their own colony. These two species are closely related honeypot ants that, like so many other ants, gather sweet plant exudates and honeydew from aphids, but, unlike other ants, store their honey for long periods of time. Lacking other containers, they have a special caste of workers, called repletes, or honeypots, that swallow the honey and store it in their own bodies as they hang quietly from the ceilings of their underground nest until the honey is needed. The honeypots become so grossly swollen that their abdomens may be almost as large as a cherry.

Slave raids by honeypot ants arise from territorial conflicts between adjacent colonies. These conflicts rarely involve deadly physical fights, as do conflicts between most other ants. They are really ritualized tournaments in which almost no ants are injured, but whose outcome leaves no doubt as to which colony is the larger and stronger. Bert Hölldobler described a tournament and its outcome in a 1976 issue of *Science*. Since it would be difficult to improve on his prose, I quote:

> During the contests the ants walk on stilt legs while raising the abdomen and head. When two hostile workers meet, they initially turn to confront each other head-on . . . Subsequently they engage in a more prolonged lateral display during which they raise the abdomen even higher and bend it toward the opponent. Simultaneously, they drum intensively with their antennae on each other's abdomen . . . This is almost the only physical contact, although each ant seems to push sideways as if to dislodge its opponent. After 10 to 30 seconds, one of the ants usually yields and the encounter ends. The ants continue to move on stilt legs, quickly meet other opponents, and the whole ceremony is repeated.
>
> I next investigated how the tournaments arise. When foragers venture into another territory, they frequently encounter foreign ants, whereupon they invariably begin to display on stilt legs. Subsequently some scouts return to their colony, [marking a chemical trail by] dragging their abdominal tips over the ground. Upon arriving at the nest, they perform a conspicuous . . . display in which they rush at nestmates over short distances and perform rapid jerking movements . . . Within a few minutes, a group of 100 to 200 ants moves out and progresses rapidly in the direction from which the scouts approached the nest . . . Upon encountering foreign conspecific workers [individuals of the same species] at the disputed territorial area, the ants invariably perform the display behavior . . . During the course of the tournament, scouts of both colonies repeatedly return to their nests and recruit reinforcements to the battleground. However, if the defending colony is considerably weaker and therefore unable to recruit a large enough worker force to the tournament area, the colony

will be overrun by the intruders and raided . . . during these raids the queens were killed or driven off. The larvae, pupae, callow workers, and honeypots were carried or dragged to the nest of the raiders. This process required several days and terminated only when the raided colony ceased to exist.

As Hölldobler and Wilson put it in *The Ants*, "the pinnacle [or nadir if you prefer] of the slave-holding way of life is reached" by the amazon ants of Europe and North America. Unlike honeypot ants, amazons do not enslave members of their own species; instead they enslave other species, on which they are totally dependent. In his classic 1910 work, *Ants, Their Structure, Development, and Behavior*, William Morton Wheeler reported that the mandibles of amazon ants are sickle-shaped weapons that are admirably suited to piercing the armor of other ants, but are not suited for digging or for other household chores. Consequently, these ants never excavate nests or care for their own offspring, and they are even incapable of obtaining their own food. Wheeler says of the amazons:

> For the essentials of food, lodging and education they are wholly dependent on the slaves hatched from the worker cocoons that they have pillaged from alien colonies. Apart from these slaves they are quite unable to live, and hence are always found in mixed colonies inhabiting nests whose architecture throughout is that of the slave species. Thus the amazons display two contrasting sets of instincts. While in the home nest they sit about in stolid idleness or pass the long hours begging the slaves for food or cleaning themselves and burnishing their ruddy armor, but when outside the nest on one of their predatory expeditions they display a dazzling courage and capacity for concerted action.

Wheeler goes on to describe a slave raid by the amazons:

> The ants leave the nest very suddenly and assemble about the entrance if they are not, as sometimes happens, pulled back and restrained by their slaves. Then they move out in a compact column with feverish haste . . . On reaching the nest to be

pillaged, they do not hesitate but pour into it at once in a body, seize the brood, rush out again and make for home. When attacked by the slave species they pierce the heads or thoraces of their opponents and often kill them in considerable numbers. The return to the nest with the booty is usually made more leisurely and in less serried ranks.

Recognizing Food

A water strider glides about in a random pattern on the calm surface of a small pond. Light enough to be supported by the thin surface film, its feet dimple the film as it skates along. In shallow water on sunny days, the enlarged, dark shadows of the dimples sweep over the sandy bottom as the insect moves on the surface with little apparent effort. Water striders have staked out their own special niche; they live by eating insects that fall onto the water. When something falls to the surface, sense organs on their legs perceive the ripples that spread in circles. Alerted to the possibility of a meal in the offing, the water strider skates to the fallen object, and, if it is an insect, it sucks the juice from its body. Although water striders are not totally indiscriminate feeders—they can, after all, choose to eat or not to eat an insect—they do not have to search far and wide for their food as do so many other insects.

The same can be said of a spider that catches flies in a web that it spins across an opening in the foliage, or of a caddisfly larva that traps food particles in a silken net that it stretches between two pebbles in a stream. Food also comes to the antlion larva that waits for unwary insects to stumble into its funnel-shaped pit in the dry sand, and to the aquatic mosquito larva that hangs below the water surface as it uses its fanlike mouthparts to make a current from which it strains edible particles.

But many insects do have to search far and wide for their food. Some female mosquitoes must find a human or some other animal from which they can suck the blood that provides the

protein that their developing eggs require. A female cicada killer searches the trees for cicadas to stock her nest. Honey bees search for blossoms. A female moth or butterfly must, before she can lay her eggs, locate a plant of the right species, one that her offspring are adapted to eat. How do these insects find the right food?

How do they recognize it—especially if, like female mosquitoes, they have never seen it before? The answers to these questions are important from an ecological point of view because they are fundamental to understanding the cohesiveness of ecosystems and the structure of food chains. They are also significant for economic reasons. The most rational approach to developing insect-resistant crop plants is to understand how pest insects find and recognize their own particular host plants. Similarly, a repellant for malaria-transmitting mosquitoes can be better designed if we understand how mosquitoes find people and recognize them as sources of food.

Insects are capable of learning. That is, they can modify their behavior according to their own experience. As we have already seen, digger wasps can remember how much food each of their offspring requires. Cockroaches can learn to run simple mazes, and ants can master even more complicated mazes.

But most insect behaviors, including food finding, are wholly or to a large degree genetically programmed. The newly metamorphosed butterfly, totally lacking in experience, is already a skillful flyer; it can find a flower and drink nectar; and it has the inborn skills required to find a mate. All this despite the obvious fact that it had no opportunity to learn or practice any of these skills during its preceding life as an egg, caterpillar, and pupa. Some birds will build complex nests typical of their species although they were raised in captivity and were deprived of so much as a glimpse of a nest throughout their lives. Similarly, the recently hatched spider spins a web typical of its species although it has never before seen a web of any sort. The full-grown cecropia caterpillar, like the caterpillars of many other moths, spins a complex cocoon when it is ready to molt to the pupal stage. But it has never before so much as seen a cocoon. Although practice may improve the caterpillar's spinning skills as it goes along—which is learning, of course—the basic blueprint of the cocoon is ultimately derived from the caterpillar's genes.

Some kinds of mosquitoes do not need to take blood meals. They retain enough protein from the larval stage to produce eggs. These species, males and females alike, sip only nectar or other sugary plant exudates as a source of energy. The females of other species must drink blood in order to produce eggs, but, like the males of their own species, they also drink plant exudates. Some of them attack reptiles, but most of them prefer warm-blooded animals such as birds or mammals. Although some mosquitoes tend to feed from either birds or mammals—but usually not both—many others are not very particular about what kinds of animals they attack. The dense swarms of mosquitoes that infest the Arctic tundra during its brief summer are equally willing to take blood from birds, people, caribou, or lemmings.

Finding a blood meal is a formidable problem for a female mosquito. She is, after all, a very tiny creature. Sitting on the skin of a person, she is barely big enough to cover a small freckle. Thus hollering distance for a human is, relatively speaking, a long and arduous trip for a little mosquito. Furthermore, how does she recognize a warm-blooded animal when she finally encounters one? In all likelihood she has never before so much as glimpsed any large animal, let alone the species from which she will take her first blood meal.

A hungry female begins her search by taking flight and cruising around. She probably is not conscious of what she is looking for but she will know it when she finds it. The female mosquito can find and identify her victim by responding to only a few cues that she recognizes innately. First, she will fly to any large, dark object that contrasts with the background. This object might be an animal, but it could also be something inanimate. In experiments female mosquitoes have, for example, oriented to dark bowling balls. If the object gives off no additional cues, she will fly past. But if it radiates heat, as do all warm-blooded animals, she will land on it. She will even land on a bowling ball if it gives off heat. If the object of her attention also gives off carbon dioxide, as do

all animals, she will probably use her piercing beak in an attempt to probe for blood.

The mosquito's ability to find a host animal and to take a blood meal is probably innate, that is, programmed in the genetic material—an "ancient memory," as Jean Auel might put it. But it is at least theoretically possible that her host finding behavior is learned or partially learned, probably by trial and error. In the case of mosquitoes, however, there is good evidence that host-finding behavior is at least largely inherited rather than learned. Even mosquitoes that have just metamorphosed from the aquatic pupal stage to the adult stage are perfectly capable of finding a host and taking a blood meal on their very first try. They have had little previous experience as adults. They certainly have had no opportunity to learn the aerial and sensory skills involved in host finding via trial and error—they never before encountered a warm-blooded animal.

How do predators recognize their prey? Are they, like mosquitoes, guided by only a few aspects of the prey animal? Simple experiments, in which models of prey were offered to various species of predators, showed that many predaceous insects and even some vertebrate predators, especially those that are not fussy about what species of prey they take, will pursue or strike out at almost anything that moves and that is of about the right size. A cat can be enticed to attack a ball of fluff that is drawn across the floor on a string. The ball of fluff acts as a simple model of a prey animal. Praying mantises will strike at paper models that do not resemble insects at all. A toad can be tricked by rolling a ball bearing in front of it. The toad strikes out at the ball bearing with its sticky tongue, brings it back to the mouth, and swallows it. These simple responses are not as silly as they seem to be at first glance. In nature, they work almost all of the time—assuming that there is no human experimenter around to confuse the situation with models. Under natural conditions, most things that move through the visual field of a mantis, a toad, or a cat will be suitable prey if they are big enough to bother with but not too big to be intimidating. The same holds for many other predators—for the robber fly or the dragonfly that sees an insect fly past, for the dragonfly nymph or the predaceous diving beetle that sees an aquatic insect or a small fish swim by.

The praying mantises, ambushers that strike at prey with their grasping front legs, will eat many different species of insects. Susan Rilling and Kenneth Roeder of Tufts University in Medford, Massachusetts, together with Horst Mittelstaedt of the Max-Planck Institut für Verhaltens-Physiologie in Seewiesen, Germany, examined in detail the prey-taking behavior of this insect. They first showed that visual stimuli are all-important to mantises in deciding whether or not they will strike out at a potential prey with their front legs, and that chemical stimuli seem to be of little or no importance. They found that mantises are as likely to strike at a fly that they could not smell because it was behind a pane of glass as one that they could smell because it was in the same cage with them. Rilling and her coworkers next used various paper models dangled from threads to determine which of the many visual aspects of a fly are actually heeded by the mantises. Mantises at least sometimes struck at any model that was between three one-hundredths and thirty-one one-hundredths square inch in area and that was moved in a jerky manner. Even simple ovals or rectangular pieces of paper of various colors elicited a few strikes. A fly-size, oval model with two wax-paper wings was slightly more effective than a real dead fly. The most attractive model of all was a similar oval with eight wax-paper wings attached, a "supernormal" model. Although mantises struck at and grasped paper models they did not attempt to eat them. But a similar effect was seen with living flies, which the mantises did eat. A fly without legs or wings elicited only two or three strikes; one with legs but no wings was struck at almost forty times; but one with both legs and wings elicited over one thousand strikes.

Praying mantises, however, are not quite the automatons that Rilling and her colleagues made them out to be. Many years later, May R. Berenbaum, of the University of Illinois, and one of her students found that mantises can learn not to attack nasty insects. Just as monarch butterflies sequester toxins from the milkweed leaves that they eat, the warningly colored, orange and black milkweed bugs sequester toxins from the milkweed seeds from which they suck their food. A mantis that eats a milkweed bug shows obvious distress, regurgitates, and soon learns not to strike at these bugs. Berenbaum proved her point by offering toxin-free milkweed bugs to mantises. These bugs, forced to eat nontoxic sunflower seeds rather than their preferred toxic milkweed seeds, were not nasty and were freely eaten by the mantises. The mantises did not, of course, learn to reject them. One of my students, Randy W. Cohen, working with Berenbaum, found that a mantis trained to reject toxic milkweed bugs also rejected a dark-colored beetle that had been painted with orange markings to resemble a milkweed bug.

You will remember that Niko Tinbergen showed that female *Philanthus* wasps use landmarks to find their way back from a hunting trip to the burrows that they stock with prey as food for their young. Tinbergen also worked on the hunting behavior of this wasp. He found that these wasps catch and stock their nests only with honey bees. They must, then, identify their prey by odor, since, like most other insects, their vision is not acute enough to distinguish honey bees from other insects by sight alone. When a female *Philanthus*, also known as the beewolf, sees a flying insect, it hovers downwind of it to pick up its odor. If it turns out to be a honey bee, the wasp pounces on it, grasps it and stings it, and then carries it back to its nest. Tinbergen found that recently killed bees, dangled from a thread so that they appear to fly, are captured in this manner and carried back to the nest. But dead bees that had been made odorless by extraction with alcohol were pounced on but never grasped, stung, or brought back to the nest. Bee-size bits of twig induced only hovering and an occasional pounce. But if the bits of twig were scented like a bee, they were often grasped and stung but were never carried back to the nest. Thus both odor and a visual stimulus are necessary to elicit an attack. Bees hidden in foliage did not elicit so much as the initial hovering response, although their odor must have been apparent to the beewolves.

While some plant-eating insects are generalists that feed on a wide variety of unrelated plants, others are specialists that live on and eat only a few botanically related species—often only a handful of species from the same plant family. Circumscribed feeding of this sort is common among insects, but it is rare among vertebrates, although the cuddly Australian koala eats only the leaves of eucalyptus—actually, the leaves of only some of the many species of eucalyptus. The milkweed bug and the monarch and queen butterflies all lay their eggs on milkweed plants, the only food that their offspring will eat. The female elm bark beetle generally lays her eggs under the bark of sick elm trees, where the larvae tunnel through the inner bark as they feed. The botanically limited menu of many economically important plant-feeding insects is revealed by their common names: pea weevil, gladiolus thrips, tomato hornworm, peach tree borer, squash bug, striped cucumber beetle, clover root borer, and chrysanthemum gall midge, to name just a few. All of these insects feed only on their namesakes and a few related plants. Common names are occasionally misleading. Corn earworm, cotton bollworm, and tomato fruitworm are different names for the same generalist insect that feeds on many plants.

The silkworm of commerce, actually the caterpillar of the silk moth, is the most famous of the botanically restricted feeders. Silk producers throughout the world grow white mulberry trees and feed their chopped leaves to silkworms. It is sometimes said that these caterpillars eat nothing but the leaves of the white mulberry, but their tastes are not quite that limited. They will eat and survive on some other plants of the mulberry family, among them the osage orange, a native of North America, and the *Cudrania* tree, whose leaves are sometimes used as a substitute food for silkworms in China.

When laying their eggs, adult insects are usually as picky as their plant-feeding offspring. They have a similar "botanical sense" and most of them lay their eggs only on plants that their caterpillars favor as food. The silk moth is an exception. After millennia of domestication, it can no longer survive in nature, having lost many behaviors essential to life in the wild, including the ability to fly and to be discriminating about where it lays its eggs. Silk moths lay their eggs on almost any surface, including paper.

The Colorado potato beetle, as its name implies, is a narrow

specialist that lays its eggs on, and will eat the leaves of, only the white potato and a handful of other closely related plants of the nightshade family. Before the American west was settled, this beetle was almost unknown. It fed only on the wild buffalo bur, a kind of nightshade, and occurred on the eastern slopes of the Rocky Mountains, where this plant is abundant. It has since become the major insect pest of the white potato and has spread throughout most of North America and Europe. There is little doubt that it will eventually spread to all of the potato-growing areas of Eurasia. This great expansion of the Colorado potato beetle's range was made possible by the introduction of a new food plant, the white potato, into the originally circumscribed range of this insect.

White potatoes originated in South America, at high elevations in the Andes Mountains, where they were and are still grown as a major food crop from Chile north to Colombia by the Incas and other native American peoples. Originally, they grew nowhere else in the world. The early Spanish explorers sent home glowing reports of this new food, but the identity of whoever introduced potato plants into Europe is lost to history. At any rate, potatoes were grown in Europe as early as the sixteenth century. By the early nineteenth century, they had become a staple food in many European countries. By that time the Irish had become so dependent on potatoes that there was massive starvation when the crop failed due to a blight caused by a fungus. During the potato famine of 1846 to 1851, about one and a half million people in Ireland— out of a population of only eight million—starved to death or were so weakened by hunger that they died of disease. Almost a million more emigrated, most of them to the United States.

The early European settlers in North America brought the white potato back across the Atlantic Ocean. As the pioneers moved west across the continent, they planted potatoes wherever they settled. In the 1850s they reached the eastern slopes of the Rocky Mountains in Colorado. After the settlers planted potatoes there, some of the beetles, which soon became known as Colorado potato beetles, left their native host plant, the buffalo bur (*Solanum rostratum*), to feed on the leaves of its newly introduced close relative, the white potato (*Solanum tuberosum*). Once this insect had accepted the white potato as a food plant, it was possible for it to do something that had previously been impossible for these beetles. It migrated eastward along a broad highway of potato patches, moving from patch to patch at an average rate of about eighty-five miles per year. It reached Illinois in 1864, Ohio in 1869, and the Atlantic coast in 1874. Shortly after World War I it appeared in Europe, and it now continues to move eastward across Eurasia wherever potatoes are grown.

Extraordinary reciprocal relationships have evolved between certain plants and certain plant-feeding insects. In some instances, plant and insect have become so interdependent that neither one could exist without the other. For example, there are about forty species of yuccas, including the large joshua tree of the southwestern United States, and recent studies suggest that each one of them is pollinated only by females of a particular species of yucca moth. During their caterpillar stage, these moths are completely dependent upon the developing seeds of yucca for food. When a female yucca moth is ready to lay her eggs, she first flies to one of the large white blossoms of a yucca and forms in her mouthparts a large ball of pollen that she gathers from the anthers, the male organs of the blossom. Then she flies to another blossom and lays one or more eggs in the ovary and carefully packs some of the pollen that she previously gathered onto the stigma, the female organ of the blossom. She repeats this operation several times with other blossoms. If she does not fertilize the blossoms, there will be no developing seeds for her larval offspring to eat. But her larvae eat only some of the seeds, leaving plenty of others to propagate the yucca.

In 1965, while collecting insects in southern Mexico, I was attacked by scores of small stinging ants when I brushed against a shrub that was swarming with them. After that, I avoided these shrubs, bull's-horn acacias. All of them were guarded by the same kind of ant. At that time, the complexity of the relationship between ants and the acacias was not fully understood. The geologist Thomas Belt had given the first account of this relationship in 1888: "These ants form a most efficient standing army for the plant, which prevents not only the mammalia from browsing on the leaves, but delivers it from the attacks of a much more dangerous enemy—the leaf-cutting ants. For these services the ants are not only securely housed by the plant, but are provided with a bountiful supply of food, and to secure their attendance at the right time and place, the food is so arranged and distributed as to effect that object with wonderful perfection."

But it was not until 1966 that Daniel H. Janzen, now a professor at the University of Pennsylvania, gave the first full account of this amazing ant-plant relationship. As Belt had discovered, the bull's-horn acacia provides food and housing for the ants, but Janzen showed that these ants cannot survive unless they are associated with this acacia, and that the acacia does not survive unless it is occupied by ants. Housing for the ants is provided by the large, swollen thorns that give the bull's-horn acacia its name. When a queen ant founds her colony, she makes her nest in a hollowed-out thorn. She gnaws a small entrance hole in the thorn, removes the pith from the inside, and then commences to lay eggs. Her colony eventually spreads to all of the thorns on the shrub and sometimes even to nearby shrubs. Sugary, high-energy food for the ants is provided by nectaries that are not associated with flowers, one near the base of the stem of each leaf. Oil and protein are provided by unique structures, called Beltian bodies, that grow at the tip of each of the many leaflets of the acacia's compound leaves.

In return, the ants guard the bull's-horn acacia. They sting and drive away cattle, deer, and mice that might browse on its leaves; they kill all insects that land on it; and they prune away all other plants that grow within several feet of their home shrub. Janzen showed that bull's-horn acacias do not survive unless they are defended by ants. Plants from which he removed the ants by spraying with an insecticide or by removing all of the thorns were heavily attacked by insects and were soon crowded by other plants.

Insects whose menus include only one plant species or a few related plants have been called the botanists or plant taxonomists of the insect world, because they make their choices according to the taxonomy, or botanical classification, of plants. How do these insect botanists pick out their own specific food plants from among so many others? How do feeding nymphs, larvae, or egg-laying females recognize familial relationships among plants? Human botanists require years of training to accomplish these tasks. The answers are complex and not yet fully understood. In the broadest sense, insects use a combination of visual, tactile, taste, and odor stimuli to recognize their food plants, but odor and taste play special and important roles. The plant-finding behavior of insects is mainly programmed in the genes. In only a very few instances is it known to be modified by learning.

Leaf-sucking insects in search of food, such as whiteflies and aphids, land on green or yellow surfaces, leaves or even pieces of paper, but they attempt to feed only on leaves and will continue to feed on them only if they smell and taste "right." Tactile stimuli may dissuade an insect from feeding, although the sight, taste, and smell of the plant may be otherwise acceptable. For example, in the midwest, soybeans with hairy leaves flourish and grow to be thirty-six or more inches tall. But soybeans with smooth leaves seldom grow to be more than eight inches tall, because they are attacked by leafhoppers, tiny insects that suck the sap from their leaves. Leafhoppers cannot penetrate down to the surface of a hairy leaf, but they can successfully feed from a hairy leaf after it has been shaved by an experimenter. Needless to say, the commercially grown varieties of soybean all have hairy leaves.

The characteristic smell or taste of a plant, either to an insect or to a person, depends upon its chemical composition. Broadly speaking, plants contain two categories of chemical substances: primary and secondary. The primary substances, such as proteins, carbohydrates, vitamins, and plant hormones, are required for growth. They are found in all plants, and no plant could survive without them. The secondary substances are a diverse and multitudinous array of chemicals that have no known role in the growth or metabolism of plants. Plants could, at least theoretically, get along without them. There are tens of thousands of secondary substances. Only a few of them occur in any one species of plant, but the same or similar ones tend to occur in related plants such

The long and slender proboscis of a sphinx moth probes for nectar deep within the flower of a trumpet creeper

as members of a single family. It is these secondary substances that give plants their distinctive tastes and smells. Vegetables and fruits such as celery, fennel, cherries, and cabbage owe their characteristic odors and flavors to secondary substances. We value spices such as vanilla and pepper and herbs such as sage, rosemary, and thyme for the odors and flavors of the secondary substances that they contain. These secondary substances give the plant the "right" or the "wrong" odor or flavor that insects use as a cue in deciding whether or not to eat the plant or to lay an egg on it.

How did this multitude of secondary plant substances that exists today come into being? They first appeared, and new ones surely continue to appear, as the result of genetic mutations in individual plants. But if a mutation is to survive and spread to the descendants of a mutant, it must pass the muster of natural selection. In other words, it must prove to be useful. It must increase the fitness of the individual plants or animals that harbor it. Some secondary substances are favored by natural selection because they are scents that attract pollinating insects to blossoms, and the floral scents that delight our noses may be even more pleasant to the insects that are rewarded with nectar when they visit these blossoms. Other secondary substances that arose by mutation were conserved by natural selection because they proved to be biochemical defenses against the enemies of plants. Viruses, bacteria, fungi, and various vertebrates are among these enemies, but I will focus on the defenses against plant-feeding insects. Some of these defensive substances are nasty, causing insects or other plant feeders to suffer unpleasant symptoms or even to die. Other secondary substances are not in themselves nasty, but are characteristic scents or tastes that dissuade the insect from feeding by warning it of the presence of some other substance that is nasty.

For hundreds of millions of years there has been an escalating evolutionary arms race between plants and plant-eating insects. This race goes on today and will continue as long as insects and plants survive on the earth. Plants evolve defenses against the insects that eat them. If the insects are to survive, they must switch to other foods or evolve ways to circumvent the plants' defenses. They may evolve a way to detoxify a nasty substance, to store it in their bodies out of harm's way, or to avoid its effects in some other manner. Insects soon come to prefer the plants whose defenses they can circumvent, and they eventually evolve the ability to identify them by their characteristic flavors or odors, or both. A plant species whose defense has been broached must then evolve another defense or fall by the wayside. The insects must respond in turn. Today we see around us the survivors of this arms race. Thousands or even hundreds of thousands of plant and animal species must have gone extinct because they could not keep up in the race.

The secondary plant substances that exist today evolved in this manner during the past 300 million years or more. As the arms race progressed, fewer and fewer plants remained as suitable hosts for any one species of insect. Insects thus tended to become associated with narrowly defined and often botanically restricted groups of plants, perhaps the members of only one genus or

family. This tendency toward botanically defined specialization is possible and persists because related plants generally contain the same or related secondary substances. Many insects eventually became so adept at eating and living on similar and related plants that it was advantageous for them to evolve as restricted specialists that associate only with these few plants and avoid all others. As the late Gottfried S. Fraenkel of the University of Illinois pointed out, some of the secondary substances that give plants their characteristic odors or flavors were adopted by specialist insects as token stimuli (sign stimuli) that enable them to recognize the plants on which they grow and survive best. Other plants are avoided because they lack "token stimuli" or because they contain deterrent secondary substances.

The common cabbage white, a small white butterfly with a few black spots, is commonly seen flying around roadsides and gardens. The caterpillars of this butterfly feed on the leaves of broccoli, cabbage, turnips, and other members of the mustard family that contain mustard oil glucosides. (These glucosides are the parent compounds of the mustard oils that elicit the warm, tingling sensation caused by mustard plasters and that give the prepared mustard on our sandwiches its biting and pungent taste.) But there are revealing exceptions. Cabbage white caterpillars also eat the leaves of two plants that do not belong to the mustard family but that do happen to contain mustard oil glucosides: the spider flower, *Cleome,* an annual garden plant that belongs to the caper family, and nasturtium, another annual garden flower that belongs to the nasturtium family. These "botanical exceptions" on the part of the cabbage white underscore the role that secondary plant substances, in this case mustard oil glucosides, play in leading insects to their proper host plants.

Shortly after World War II, Asgeir J. Thorsteinson did a series of experiments that proved to be milestones in the effort to understand how insects find and recognize the plants that they live on and eat. He was working on his doctorate under the direction of Gottfried Fraenkel, who was then at the University of London in England. He found that the diamond back moth, which also eats only plants of the mustard family as a caterpillar, can be induced to eat the leaves of several non-mustards, including pea and cucumber plants, that it does not ordinarily eat. These leaves are readily accepted if they are coated with mustard oil glucosides. Mustard-feeding insects even try to eat filter paper that has been similarly coated. These observations leave little doubt that the mustard oil glucosides are token stimuli for at least some of the insects that feed on plants of the mustard family. Thorsteinson's work is, of course, fascinating from the point of view of a naturalist, but it also has important practical implications for plant breeders who develop varieties of crop plants that are resistant to the attacks of insects.

Botanical specificity obviously sets a limit on the number of plant species that an insect can eat. The cabbage white can exploit dozens of plants but there are tens of thousands of others that it will not eat. This would seem to be a disadvantage. But botanical specificity is so common and widespread among insects that it is hard to believe that it does not confer advantages that more than make up for the loss of flexibility in feeding. It may even be that insects were not just passive pawns in the arms race. They may have cooperated in narrowing their own choice of food plants.

Like monarch butterflies, some insects turn the poisons in their host plants to good use. Many other warningly-colored insects are not affected by the poisons in their food plants but, like monarchs, sequester them in their bodies so that they themselves are toxic to animals that may prey on them. The likelihood that insects can gain other advantages not necessarily related to nutrition, the avoidance of plant toxins, or their sequestration as defenses, becomes apparent once we realize that the host plant is not simply the insect's food. It is also the place where the insect lives. By narrowing its host-plant range to only certain species, an insect may avoid parasites or predators that are specifically associated with other plants. The host plant may provide a favorable microclimate and it certainly provides cover and background for the creatures that live on it.

It is obvious to the eye that specialists are often better camouflaged on their usual host plants than they would be on others. Many caterpillars and sawfly larvae that feed only on pines are

patterned with thin, alternating stripes of dark and light green that run the length of their bodies. They are very difficult to see when they rest on pines with their bodies parallel to clusters of the thin green needles. They are much easier to see if they are transferred to broad-leaved plants. L. de Ruiter, a Dutch ethologist, pointed out how different species of stick caterpillars have become specialized in appearance to blend in with the particular host plants on which they feed. During the day, these caterpillars sit frozen in place so as to resemble a short twig jutting out from a branch. A species that lives on birch has a slender and relatively smooth body that resembles a birch twig. Another species, which lives on oak, has a thicker and bumpier body that resembles an oak twig. Both species also closely match in color and pattern the twigs of their respective host plants.

Botanical specialists avoid poisons that they cannot detoxify, but whether or not they gain nutritional advantages is an open question. Plants of all families do, after all, contain the vitamins, minerals, and other nutrients required by insects and all other animals. Whether or not they contain them in a balance that is suitable for any particular insect species is another question and one for which we have few answers. My own doctoral dissertation, done with the guidance of Gottfried Fraenkel, showed that if certain organs of taste on the mouthparts are removed from tobacco hornworms, they will obtain all of the nutrients that they need for growth by eating plants, such as dandelions, that are not related to tobacco and that they would not ordinarily eat. Some of the plants that these surgically altered tobacco hornworms ate supported only slow and inefficient growth. But it may well be that these plants were poor foods because they contained noxious secondary substances, not because they were short of nutrients.

Since prehistoric times, humans have known and have taken advantage of the medicinal, insecticidal, mind-altering, and even the poisonous properties of secondary plant substances. These properties are understood by the most primitive tribal cultures known today, and they may well have been understood by our distant ancestors even before they had evolved as full-fledged humans. It was recently discovered that wild chimpanzees, already famous for their ability to learn sign language and make simple tools, use some medicinal plants. Even starlings incorporate the green foliage of an insecticidal plant in their nests. Young starlings in nests with fleabane foliage have far fewer parasitic mites than do young birds in nests that contain no fleabane. This plant's traditional common name indicates that its antiparasite properties have been known for a long time.

Aspirin, the first miracle drug, is without doubt the most widely used medicine in North America and elsewhere. The aspirin that we use today is synthetic, but its parent compound occurs in the leaves and other parts of various plants, most notably in the bark of several species of willow. Long before the appearance of Columbus, native Americans understood the beneficial properties of willows and used decoctions of the bark, roots, or leaves to alleviate headaches and to subdue fevers. Extracts of willow were used by the ancient Greeks, and their use is prominent among the folk remedies of European and Asian peoples. Although we use many synthetics today, plants remain an important source of medicinal substances, and the majority of the synthetic medicines are just manmade copies or variants of compounds that were originally discovered in plants or other organisms.

There are hundreds or even thousands of secondary plant substances that have medicinal or other properties that people have found to be useful. Among them is digitalis, used to slow down and strengthen the heart beat; it is derived from foxglove, a garden flower of the snapdragon family. Quinine, obtained from the bark of the South American cinchona tree, is used to treat malaria. It has become less useful because some strains of the malaria-causing protozoa are now resistant to it—just as the mosquitoes that transmit these protozoa from person to person are now resistant to insecticides that once killed them. The currently most useful treatment for amoebic dysentery, emetine, is obtained from the roots of ipecac, a creeping plant of the madder family that is native to South America. Among the most useful modern medicines for treating childhood leukemia are vincristine and vinblastine, only recently discovered in the periwinkle, a garden plant of the dog-

bane family. Taxol, an antitumor agent found in the bark of yew trees (*Taxus*), was approved for use in humans in 1992. It is effective in treating ovarian cancer where other chemotherapies have failed and may be useful in treating breast cancer, lung cancer, and malignant melanoma. It takes 16,000 pounds of bark harvested from wild yews to produce 2.2 pounds of taxol. In 1993 the first commercially grown yew seedlings were harvested in the state of Washington as a source of taxol. But even more recently taxol has been synthesized in the laboratory.

Not surprisingly, many plants other than fleabane contain secondary substances that are insecticidal. Among these insecticides derived from plants are ryania, hellebore, sabadilla, rotenone, pyrethrum, and nicotine. Pyrethrum, the most widely used of the botanical insecticides, was first obtained centuries ago from the flowers of a chrysanthemum, but there are now several synthetic pyrethroids that are widely used in agriculture and as household insecticides. Rotenone does double duty as a fish poison and is currently used in North America and elsewhere to kill undesirable fish in ponds and lakes. The Indians of South America have long used it to harvest fish from rivers and other bodies of water. Cubé roots, the source of rotenone, are mashed and thrown into the water. The dead fish rise to the surface and can be safely used as food for humans.

The dark side of the story of the secondary plant substances is that humans have used some of them as poisons to kill other humans. People of different cultures, including the Chinese, the ancient Greeks, and several African and South American tribes, used various plant poisons to tip the arrows that they used in warfare. Indeed, our English word "toxic" comes to us from the Latin *toxicum*, a poison, which was derived from the Greek *toxikon*, an arrow poison. In 1964 an American missionary in Africa was killed by a poisoned arrow shot into her back. There may have been more recent killings of this sort in Africa or South America, but I have not heard about them. The Chinese tipped arrows with aconite, derived from a common garden plant, monkshood (*Aconitum napellus*), of the buttercup family.

This poison has been well known in Europe for millennia.

According to Walter Lewis and Memory Elvin-Lewis, in ancient times convicted criminals were forced to drink a fatal dose of aconite, and "on the Greek island of Ceos, infirm old men were compelled to take a draught" of it. Shakespeare knew it well. The "dram of poison" that Romeo wished for was aconite. In *King Henry IV, Part 2*, Henry compares its power as a poison to the explosive power of gunpowder: "it [the venom of suggestion] do work as strong as aconitum or rash gunpowder."

In 399 B.C., Socrates was convicted of "corruption of the young" and "neglect of the gods" by his fellow Athenians. He was sentenced to drink an extract of the fatal hemlock, a member of the parsnip family. In 1776, a petty king in Java had thirteen of his concubines put to death after they were convicted of infidelity. According to Lewis: "The girls were lashed to posts, their breasts bared, and an awllike instrument poisoned with the latex of upas lanced the unhappy women about the middle of their breasts. All died within five minutes in great agony." The latex, or milky sap, of the upas tree, a relative of the mulberry, is safe to imbibe but deadly when it enters the bloodstream.

Plant poisons have been used in trials by ordeal. Among the people who have used them in this way are the Efik of Calabar, in what is now southeastern Nigeria. The Efik, an Ibibio-speaking tribe, were originally fishermen but prospered as slave traders in the seventeenth and eighteenth centuries. They had a great fear of witchcraft and used the deadly Calabar bean (*Physotigma venenosum*) to determine the guilt or innocence of people accused of being witches who practice black magic. The suspect was made to drink ground-up Calabar beans added to water. He was judged innocent if he raised his right hand and threw up the mixture. If a person thus found to be innocent suffered symptoms of poisoning even after vomiting, he was given aid in the form of a drink that is probably a good emetic, excrement mixed in water that had been used to wash the external genitalia of a female.

An Efik man speaking pidgin English described the effect of the beans on a person who is poisoned. "Him do dis, soap come out of him mout, and all him body walk." In 1858, Her Britannic Majesty's consul for the Bight of Biafra remarked that these words are "a most perfect description of the frothing from the mouth, and the convulsive energy of the whole frame," that follow the ingestion of Calabar beans. Convulsions and the other symptoms appear because Calabar beans interfere with the normal transmission of nerve impulses. Daryll Forde's *Efik Traders of Old Calabar* and Tony Swain's *Plants in the Development of Modern Medicine* provide fascinating reading on the Efik and their use of this toxic bean.

Physostigmine, the toxic principle in Calabar beans, is poisonous to virtually all vertebrates and insects. Not surprising, since the nervous systems of these animals transmit impulses in similar ways. In 1864, Thomas R. Fraser, M.D., Assistant to the Professor of Materia Medica in the University of Edinburgh, wrote: "The Ordeal-bean of Calabar is a poison of extreme activity; hitherto no living being had been known to be able to resist its action; and, from my knowledge of its properties, I confess to having been skeptical of the existence of any animal form which could be fairly subjected to its influence and still retain its hold on life."

The exceptional animal that is not poisoned by Calabar beans is the caterpillar of a tiger moth that was sent to Fraser by the Reverend John Baillie of Old Calabar. The package that Fraser received from Baillie contained Calabar beans, cocoons, caterpillars, and their excrement webbed together in a tangle of silken strands. The caterpillars in the package were eating the beans and continued to do so until they were ready to become pupae. The astonished Fraser tested the caterpillar excrement on a bird and found it to be highly toxic: "Half a grain of the detached cylinders of excrement was triturated [ground up], moistened, and formed into a small pill, which was placed in the pharynx of a linnet. Perfect paralysis of the legs was caused in four minutes, together with marked contraction of the pupils . . . defaecation and lachrymation. In seven minutes, life was extinct."

When Fraser used this unfortunate bird to test for the presence of the toxin, there was no chemical means to identify this toxin because it had not yet been isolated and chemically described. Shortly thereafter he isolated an almost pure form of the toxin from the beans and the excrement of the caterpillars. He named the toxin eserine from the Efik word for the Calabar bean, *eséré*. Just a year later, in 1864, other pharmacologists purified it completely and named it physostigmine. Both names are still used today.

The adaptation that allows the caterpillar to eat Calabar beans and excrete the toxin without itself being harmed is a notable episode in the escalating arms race between insects and plants. There is no telling what the situation will be thousands of years in the future, but right now the Calabar bean is still ahead of the game. Only a few insects can eat its seeds and survive. Indeed, chemical compounds, carbamates, that are based on physostigmine are now among our most important insecticides—largely due to the pioneering efforts of Robert L. Metcalf, one of my colleagues at the University of Illinois.

Many plants contain secondary substances that have hallucinogenic or other mind-altering properties. Among them are coffee, tea, mescal, peyote, marijuana, coca, and opium. Peyote, native to the Rio Grande valley of Texas and to parts of the Mexican plateau, is a small and unobtrusive cactus that contains a powerful hallucinogen. Native Americans, including the Aztecs, used it in religious ceremonies and continue to do so to this day. The natives of the Andean highlands of South America chew coca leaves as a stimulant and a hunger depressant. Coca-Cola contained small amounts of coca until its use for nonmedicinal purposes was prohibited in 1904. When coca is taken in such dilute and unpurified doses, its effect is reasonably benign. But when coca's active ingredient, cocaine, is refined and taken in large doses it can have disastrous effects and can be highly addictive. An Andean Indian chews about two ounces of dry coca leaves per day, which yields a total dose of about 46 milligrams of actual cocaine. But the average cocaine addict takes a daily dose of about 467 milligrams of pure cocaine, only a little less than half of the lethal dose.

The fly agaric, *Amanita muscaria*, a fairly common mushroom in the northern forests of Eurasia and North America, figures in one of the most bizarre stories in the annals of hallucinogenic drugs. Deadly only in large doses, this mushroom is related to several much more toxic species, including the death cap or deadly angel, *Amanita phalloides,* probably the most deadly of all the fungi. Several Siberian tribes use the fly agaric as a hallucinogen. R. Gordon Wasson, in his *Soma, Divine Mushroom of Immortality,* quotes a 1736 report by Filip Johann von Strahlenberg, a Swedish colonel who spent twelve years as a prisoner of war in Siberia and made observations on the Koryak tribe's use of the fly agaric:

> The Russians who trade with them [the Koryak], carry thither a kind of mushroom called, in the Russian tongue, Muchumor, which they exchange for squirrels, fox, hermin [sic], sable, and other furs: Those who are rich among them, lay up large provisions of these mushrooms, for the winter. When they make a feast, they pour water upon some of these mushrooms, and boil them. They then drink the liquor, which intoxicates them; The poorer sort, who cannot afford to lay in a store of these mushrooms, post themselves, on these occasions, round the huts of the rich, and watch the opportunity of the guests coming down to make water; And then hold a wooden bowl to receive the urine, which they drink off greedily, as having still some virtue of the mushroom in it, and by this way they also get drunk.

Wasson goes on to quote from a 1774 report by Wilhelm Georg Steller on the Kamchatka Peninsula of Siberia:

> In the Russian settlements this habit [using fly agaric as a hallucinogen] has been lost for a long time. However, around the Tigil and towards the Koryak border it is very much alive. The fly-agarics are dried, then eaten in large pieces without chewing them, washing them down with cold water. After about half an hour the person becomes completely intoxicated and experiences extraordinary visions. The Koryak and Yukagir are even fonder of this mushroom. So eager are they to get it that they buy it from the Russians wherever and whenever possible. Those who cannot afford the fairly high price drink the urine of those who have eaten it, whereupon they become as intoxicated, if not more so. The urine seems to be more

powerful than the mushroom, and its effect may last through the fourth or the fifth man.

Wasson argues that soma, the deified hallucinogen of the ancient Veda of India, is actually the fly agaric. These people spoke an Indo-European language, Vedic, a parent language of Sanskrit. They came to India from the west, probably from the Iranian area, over 3,000 years ago. He quotes from the *Rig Veda*, a compilation of over a thousand Vedic hymns and mantras: "Soma, storm cloud imbued with life, . . . navel of the way, immortal principle, he sprang into life in the far distance. Acting in concert, those charged with the office, richly gifted, do full honor to soma. The swollen men piss the flowing [soma]."

According to several sources, members of the North American "counter-culture" have been experimenting with fly agaric as a hallucinogen. But I have not found out if they drink each other's urine.

Whether or not mood-altering hallucinogenic substances render a service to the plants that contain them remains an open question. Perhaps they help to protect plants against plant-feeding insects. Could it be that hallucinogens make insects so "high" that they wander off and forget all about feeding on and thus injuring the plants that contain these substances? This idea is not too far-fetched. Consider the experimental results obtained by Peter Witt, a Swiss pharmacologist who fed mind-altering substances to web-spinning spiders. The effects of caffeine and LSD (lysergic acid), for example, could be seen in the webs spun by dosed spiders—and the web is, after all, the product of whatever mind a spider has. Spiders on LSD spun abnormally perfect webs, whose construction was far neater and probably more time-consuming and wasteful of energy than is necessary to catch flies. Spiders on caffeine seemed to suffer from "coffee nerves." They made webs that were nothing more than a loose and haphazard tangle of threads that would catch very few flies.

Taking Nourishment

In 1940, Victor Lindlahr, a popular health food advocate of the day, published a book called *You Are What You Eat*. There is some truth in this title in the short-term, day-to-day sense. Our bodies are composed of molecules that we eat; if we eat a good diet, we will probably be healthy, but if we eat a poor diet, we will eventually become sick. In 1980, Thelma Barer-Stein turned the tables by publishing *You Eat What You Are*. She wrote about ethnic foods and the long-term impact of cultural traditions on Canadian diets. But her title could serve as well for a treatise on the effect of an animal's diet on its evolutionary heritage. The behavior and anatomy of any animal are shaped by evolution to serve, in part, for procuring and processing the particular foods that it eats. The short neck and grinding molars of the bison make for the efficient cropping and chewing of grass, just as the long neck and similar molars of the giraffe serve for browsing on the leaves of tall trees. The anteater's long snout and even longer, sticky tongue are suited to its diet. The swift flight, sharp talons, and hooked beak of the peregrine falcon are adapted for catching and eating swiftly flying birds.

And so it is with insects. How would a grasshopper or a cecropia caterpillar chew leaves without its stout, grinding mandibles? How could bed bugs, mosquitoes, and fleas drink blood if their mandibles were not drawn out to long, bladelike stylets that form part of their long, thin, piercing beaks? Black flies in the North Woods could not get at our blood if their broad, serrated mandibles, overlapping like the blades of a pair of scissors, were not there to snip holes in our skin. The praying mantis, lying in ambush, would catch very few insects if its front legs were not so nicely adapted for snatching and grabbing. How would butterflies and moths sip nectar from deep flowers without their long soda-straw mouthparts? How could they fly if the soda-straw tongues, often as long as the body, did not coil up beneath their heads when they are not drinking?

The answer to these questions is that no insect or other animal could eat as it does if it did not have the anatomical and behavioral capabilities that it has. Evolution might, of course, have shaped the animal in some other way to attain the same dietary end. There is more than one way to skin a cat. After all, both the mosquitoes and the black flies in our northern forests drink the blood of humans and moose. They just get at it in different ways. Aphids and thrips both eat the juices of plants. But the aphid uses its piercing beak to suck from beneath the surface of a leaf, while the thrips uses its differently formed mouthparts to rasp away the surface of a leaf and thus expose its juices. But evolution molded animals to be as we find them, and understanding how they use their anatomical equipment is not just an interesting intellectual challenge. Knowing how insects take their food is important to understanding ecosystems. For eating each other is one of the most important ways in which organisms interact—including, as you will read below, ways in which insects and people interact.

People do eat insects. North Americans have been known to eat chocolate covered ants. Some Africans savor termites, especially the huge, egg-filled bodies of queens; people of the Near East and Asia eat grasshoppers, and Mexicans eat the eggs of an aquatic bug, and put an agave worm, actually a chubby caterpillar, into each bottle of mescal—and two of these caterpillars into each bottle of a brand of tequila called Dos Gusanos, which means two caterpillars. But in the great scheme of things, insectivory by humans has only a negligible effect on insects. By contrast, the feeding habits of insects have important effects on humans. Blood-feeding female mosquitoes transmit the protozoa that cause human malaria. Other blood-feeding insects transmit various protozoa, bacteria, and viruses that cause other diseases in humans and our domestic animals. Worldwide, at least 10,000 species of insects eat our crops or transmit diseases to them. Boll weevils of cotton destroy the pods that produce the cotton fiber; Hessian flies stunt the growth of wheat; and the burrowing of corn borers weakens the stems of one of our most important crop plants.

As I already pointed out, adult and immature insects with gradual metamorphosis have mouthparts of the same type. Regardless of age, grasshoppers use their strong jaws to masticate the leaves of plants. The nymphs and adults of other insects with gradual metamorphosis differ only slightly in the structure of their mouthparts or in the type of food that they eat. Termites, crickets, and katydids of all ages chew on various plant parts, and cockroaches chew their omnivorous diets. Both nymphal and adult chewing lice use their mouthparts to ingest bits of skin or other tissues from the birds or mammals that they parasitize. Other insects with gradual metamorphosis have, in both the nymphal and adult stages, lengthened mouthparts that are modified for piercing and sucking. Cicadas, aphids, leafhoppers, and many others pierce plants to drink sap. Ambush bugs and assassin bugs suck the body juices of other insects. Sucking lice pierce the skin of mammals to obtain blood, and some true bugs suck blood from either mammals or birds.

The mouthparts of both nymphal and adult aphids form a long, flexible beak that is used to pierce plants to obtain their juices. Many aphids probe about within the plant's tissues and insert the beak into one of the tiny phloem tubes. Sap flows downward through these tubes, carrying nutrients manufactured in the leaves to the stems and roots. Aphids can control the rate at which they take in phloem sap, but they do not actively suck it up. It is, rather, forced up into their beaks because it is under pressure in the phloem tubes. J. S. Kennedy and T. E. Mittler discovered that, when you snip a feeding aphid away from the plant, leaving its beak embedded, sap continues to flow from the severed beak, although the sucking organs are no longer present. The chemical nature of phloem sap is a central problem of plant physiology, but it was this entomological discovery that first enabled plant physiologists to easily obtain samples of pure phloem sap.

Phloem sap is watery and rich in sugars, but is usually low in other nutrients. Consequently, aphids take in more than enough water and, except for a few species, an excess of sugar too. The excrement of aphids is thus little more than a solution of sugars in water—hence the name honeydew. Many aphids produce several times their own body weight of honeydew every day, some excreting a droplet as often as once every ten or twenty minutes. They kick away each droplet with their hind legs as it is expelled from the anus, thus preventing the sugary honeydew from encrusting and gumming up their bodies.

Since aphids often occur in large numbers and infest many different kinds of plants, honeydew is abundant and ubiquitous, both in natural situations and in our gardens. Leave your chrysanthemums untended, and they may soon teem with colonies of aphids that glaze the leaves with honeydew. Stand under an infested maple tree, and you will feel the fine mist of honeydew that constantly rains down. Examine almost any leaf, perhaps one from a redbud, a tulip tree, a sycamore, or an oak. You will probably find glistening spots of dry honeydew dotting its upper surface. Honeydew also falls to the ground and to twigs, branches, and the trunks of trees. It often persists through the winter, especially on dead oak leaves that cling to the tree.

Honeydew is so universally available and abundant that many insects—and even humans—regularly use it as food. It is an important source of sugar for wasps and bees. Honey bees collect large quantities and convert it to honey. The Germans prize honey made from honeydew that bees collect from the firs in the Black Forest. The mouthparts of many of the 90,000 known kinds of flies are adapted for eating dried drops of honeydew. Among them are hover flies, fruit flies, house flies, and even blood-sucking horse flies. Ants are doubtlessly the most famous of the honeydew collectors. Some collect honeydew where and when they can find it, making little or no effort to protect the aphids. Others guard aphid colonies that they happen to find, protecting them from parasites and predators. Some ants build shelters over aphid colonies, and some, such as the North American cornfield ant, keep aphids as permanent domestic animals. Corn root aphids do not survive for long unless they are tended by cornfield ants. During the winter, the ants store the aphids' eggs in their underground nests. When the eggs hatch in spring, the ants place the nymphal aphids on the roots of corn or other grasses, where they feed and produce the honeydew that will be harvested by the ants.

In Hebrew, honeydew is known as *man*. In Arabic, both aphids and honeydew are known as *man*, and *man-es-simma* is the honeydew, or manna, that falls from the sky. In his *Insects as Human Food*, the Israeli entomologist Friedrich S. Bodenheimer proposed that honeydew is the manna from heaven that sustained the ancient Israelites as they crossed the Sinai Desert in their flight from Egypt. To this day, honeydew is collected for human consumption in Near Eastern countries, including Turkey, Iraq, and Iran.

In Turkey and in northern Iraq, the Kurds collect huge quantities of honeydew, manna, from oak trees infested by aphids—at least they did before the Gulf War that followed Iraq's invasion of Kuwait. Branches are cut from the trees in the early morning before the ants can gather the manna. The branches are beaten to knock off the manna. In the dry air of this region, it soon hardens into a rocklike mass containing some aphids and fragments of oak leaves. Bodenheimer reports that in the 1940s the annual production of manna in this area was about 70,000 pounds. In its raw state, the manna is sold to confectioners, who dissolve it in water and strain it through cloth to remove contaminants. According to Bodenheimer, the purified honeydew is then mixed with almonds, seasonings, and eggs, about fifty eggs for each three and a half ounces of honeydew. This mixture is boiled and allowed to solidify, after which it is cut into pieces and coated with powdered sugar. It sounds delicious, but unfortunately I have never had a chance to taste it. However, an Iraqi acquaintance told me that it is indeed delicious, and that, once tasted, its flavor is never forgotten.

The sucking lice, another group of insects with gradual metamorphosis, pierce the skin of humans and other mammals in order to suck their blood. They do not live on koalas, opossums, or other marsupials, nor do they occur on the monotremes, the egg-laying mammals: the echidna and the duck-billed platypus. But they do live on many other mammals, including cattle, giraffes, camels, peccaries, goats, mice, squirrels, wolves, wild cats, seals, and walruses. Sucking lice tend to be fussy about the mammals on which they live. Some live on only one species, and closely related species of lice usually live on closely related species of mammals. The crab louse, the body louse, and the head louse, for example, live only on humans, and all of their close relatives live on monkeys or apes.

Sucking lice are so completely parasitic that they often spend their entire lives on one host. A young louse may grow to maturity in the soft fur of a mouse, court and copulate there, and, if it is a female, glue its nits, or eggs, to the hairs. Long ago in evolutionary time, the lice lost their wings as an accommodation to living in dense hair or fur, and today they can switch hosts only when their own host is in close bodily contact with another. The crab louse, also known as the pubic louse, moves from person to person mainly during sexual contact. Other lice move from parent to offspring or from litter mate to litter mate. Sucking lice are so tenacious in holding on that they are seldom torn loose by scratching and almost never fall off their host as long as it is alive. They press their flattened bodies down against the host's skin while each of their six legs, specially modified for grasping, holds tight to the base of one of the host's hairs.

Sucking lice eat nothing but blood; blood, however, is a poor source of the B-vitamins that are required by insects and all other animals for growth, reproduction, and their very existence. How, then, do blood-sucking lice survive? As pointed out by Sir Vincent B. Wigglesworth, lice, other blood-sucking insects, and insects that eat such vitamin-poor foods as wood or plant sap (aphids, for instance) harbor in their bodies friendly microorganisms, possibly bacteria or fungi, that supply them with the missing vitamins. It is a fair trade, a mutualistic relationship. The microorganisms get a favorable place to live, and the insects get the vitamins that they need. In nymphal lice, these microorganisms are housed in special organs associated with the digestive system. The microorganisms

degenerate in adult male lice, but in females they migrate to the ovaries, where they are incorporated in the eggs. They are thus passed from generation to generation.

The body lice of humans, known as "cooties" during World War I and often called mechanized dandruff by the GIs of World War II, may transmit several diseases when they take their blood meals. By far the most important of them is typhus fever, which is debilitating and often deadly. The mortality rate is low in small children but increases with age, ranging from 10 to 100 percent in adults of different ages. Lice acquire the microorganisms that cause typhus fever when they take blood from an infected person. These pathogens multiply enormously in the digestive system of the louse and cause its death within a few days, but in the meantime the louse continues to feed. Infected lice that suck blood from people defecate near the site of the puncture. The person becomes infected when he or she scratches the louse bite, rubbing the louse feces, which teem with the typhus-causing microorganisms, into the puncture wound.

Historically, typhus fever, a disease of cool climates, has been particularly important in Europe. Great outbreaks of body lice and typhus fever occurred among soldiers in the field, prisoners in concentration camps, residents of bombed-out cities, and others who could not keep their clothing clean. Under such conditions, a person's clothing may be infested with hundreds or even thousands of lice, and when people are crowded together the lice easily move from person to person. The armies of Napoleon, Kaiser Wilhelm, and Hitler all suffered outbreaks of typhus fever. Napoleon's ill-fated invasion of Russia began with an army of 453,000 men in late June of 1812. When his defeated army left Russia in early December of the same year, only 20,000 sick and discouraged men remained. As Hans Zinsser pointed out in his *Rats, Lice, and History*, Napoleon's catastrophic losses were due more to rampant disease, especially louse-borne typhus, than to enemy action or the fierce Russian winter. The last great epidemic of typhus fever in Europe occurred in Naples in 1943 and 1944 during World War II. The allies stopped the epidemic by dusting about three million people, civilians and occupying troops alike, with DDT and other insecticides.

Bed bugs feed only on blood, both in their nymphal and adult stages. They are notorious for taking blood meals from humans, but they also attack chickens and bats. Relatives of the bedbug drink the blood of many different kinds of bats and of swallows, chimney swifts, and other kinds of birds. Like the sucking lice, bed bugs and their relatives have internal symbionts that supply B-vitamins and that can be passed from mother to offspring through the egg. But, unlike the lice, bed bugs are not known to transmit diseases of humans, although they have been suspected of doing so. Nor do bed bugs live permanently on humans or other hosts. In human dwellings, they spend the day hiding under bedding, in cracks or crevices, or under loose wallpaper, but at night they come out to suck blood from sleeping humans.

Bed bugs are now uncommon in the United States, but in the 1940s, when I was growing up on the wrong side of the tracks in Bridgeport, Connecticut, they still infested many tenements. If you were hunting for a new apartment, one of the first things that you did was to search for bed bugs, peering into cracks and crevices and lifting up pieces of loose wallpaper. In those days you could still see people in their backyards trying to get rid of bed bugs by flaming their metal bedsteads and bedsprings with kerosene. Bed bugs occasionally infested the seats in theaters, trolley cars, or trains, sucking blood especially from the bare or thinly clad legs of women. In Bridgeport in those days there was one movie theater that was still known as the bug house, but I was never bitten by a bed bug when I went there. In some areas of the world, bed bugs still infest public transportation. I clipped the following letter to the editor by Mohammad Khalid Malik of Lahore from the English-language *Pakistan Times* of Monday, May 20, 1985:

> I would like to invite the attention of the Railway authorities towards the sad plight of express trains running between Karachi and Peshawar.

These trains are infested with bugs, if a white piece of cloth is spread on the berth, it turns red owing to these bugs. As such, most of the passengers prefer to sleep on the floor. But the bugs start falling on the floor from the berth just like rain-drops seeping through the roof of a poor man's cottage.

On being tortured by the bug bite, I started killing the insects. A shrewd passenger said to me, "Why are you killing these bugs?" I retorted: "They have sucked my blood." The man said, "Well, human blood is an object of love for these bugs and they are thus our kids." I was constrained to deduce that this very idea is, perhaps, preventing the Railway authorities from taking any remedial measure to eliminate the bug menace.

But one thing is certain: if these bugs are allowed to multiply, a "bug special" will have to be run instead of a passenger train.

Although the bed bug now occurs in cool climates all over the world, the late Robert L. Usinger of the University of California at Berkeley speculated that they originally occurred only in the Old World and that they attacked only bats, but that they became associated with humans when our distant ancestors in the Middle East and Europe first moved into caves that were occupied by bats. In his book on bed bugs, Usinger presented linguistic evidence that virtually proves that bed bugs originated in the Eastern Hemisphere and were unknown in the Western Hemisphere before European colonization. There is no word for bed bug in any of the native American Indian languages. But all Indo-European, African, and Oriental languages that have been checked do have a word for bed bug. In Italian the word is *cimice*, derived from the Latin *cimex*, from which is derived the scientific name for the bed bug family, Cimicidae. In German it is *Wandlaus*, wall louse, and there are related words with the same literal meaning in other Germanic languages: *wandluis* in Dutch, *vägglus* in Swedish, and *vaeggelus* in Danish. In Polish the word is *pluskwa*, in Hebrew it is *pishpesh*, and in Douala, a Bantu language of southern Africa, it is *ekukulan*.

There are more differences between dragonfly nymphs and adults than there are between the nymphal and adult forms of most other insects with gradual metamorphosis. The nymphs are aquatic and the adults are aerial. Both nymphal and adult dragonflies have several peculiarities. The nymphs use gills that line the inside of the rectum to breathe oxygen dissolved in water. Like a rubber-bulb syringe, the muscular rectum sucks fresh water in through the anus and expels it by the same route after it has bathed the rectal gills. It can be expelled forcefully enough to propel the nymph across an aquarium. The nymphs are voracious predators that eat other aquatic insects and even small fish. The air-breathing adults, usually seen flying back and forth over a still pond or perched on vegetation near the shoreline, are also predators, but eat flying insects such as mosquitoes, butterflies, or even smaller dragonflies of other species.

Although nymphs and adults both use their jaws to chew up prey, the nymphs have a special adaptation of their mouthparts that adults do not have. The adults, which hunt while on the wing, capture their aerial prey in a "basket" formed by their spiny legs and then reach down to grab it in their jaws. The aquatic nymphs, by contrast, ambush prey from a hiding place. Some of them bury themselves in the bottom muck with only their eyes protruding above its surface—as the eyes of a frog or an alligator protrude above the surface of the water. When a potential meal swims by, another insect or even a small fish, they strike out with their extraordinarily long and jointed lower lip, which may be half of their body length, grab the prey with a pair of fanglike teeth at the end of the lip, and then bring the prey back to their chewing jaws as they refold the enormous lower lip beneath the body.

In insects with complete metamorphosis, the feeding organs of larvae and adults may be entirely different. Caterpillars have chewing mouthparts for eating leaves, but when they become adult moths or butterflies they have siphoning mouthparts for sucking up liquids such as nectar. Adult fleas use their piercing-sucking mouthparts to take blood from mammals or birds. Flea larvae, however, have chewing mouthparts for eating organic

A nymphal dragonfly uses its long, prehensile lower lip to snatch a small fish

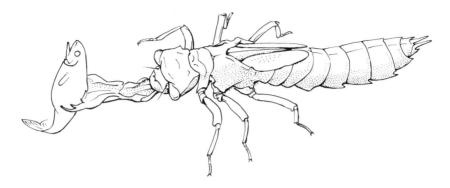

debris that they find in the lair or nest of the host mammal or bird. Adult caddisflies, commonly seen at night flying about lights near ponds or streams, have reduced mouthparts. Some species of caddisflies may not be able to feed, but others drink liquids such as nectar. The aquatic larvae have chewing mouthparts. Some species that live in rapid streams build intricate silken nets that strain their food—algae and tiny animals—out of the flowing water.

The differences between larvae and adults are less marked among beetles than among other insects with complete metamorphosis. Both grubs and adults have chewing mouthparts. Some members of the leaf beetle family, such as the Colorado potato beetle, chew leaves in both of these life stages, but northern corn rootworms eat roots as grubs and flowers as adults. Larval whirligig beetles live under water and capture living prey. Adult whirligigs skim about erratically on the surface film as they scavenge aerial insects that fall to the water. Adult blister beetles eat foliage, but the grubs eat grasshopper eggs or the stored food and eggs in the nests of solitary bees.

The larvae and adults of some beetles, including those known as water tigers and those known as fireflies, have mouthparts that are basically like those of other beetles but are adapted for sucking liquids from their prey. The mandibles, rather than being short, stout structures adapted for chewing and grinding like a molar tooth, are long, sickle-shaped, and traversed by an internal duct that runs from the pointed tip of the mandible to the mouth opening. The larva grasps and pierces its prey with its opposing mandibles; it injects the prey with digestive juices, and then it waits for these juices to act before it sucks out the liquified internal parts of its prey.

The caterpillars of moths and butterflies have chewing mouthparts. Like cecropia, most of them eat the leaves or other structures of plants, but some of them have different feeding habits. A few are predators. The caterpillar of a North American butterfly, the harvester, eats woolly aphids. It lives in aphid colonies that occur mainly on alders, and soon becomes covered with the fine waxy threads that cover the aphids and give them their name. Stored spices, herbs, grain, and bakery products are the dwelling places and the food of the caterpillars of a few species of moths. The notorious clothes moths that invade our homes eat woolens and fur coats when they are larvae. Under natural conditions, they eat the hair and dry skin that remain after maggots, vultures, and other scavengers have eaten the other parts of a dead animal.

While most plant-feeding caterpillars browse on leaves, some live in fruits, burrow in stems—even the woody trunks of trees—or chew on roots. (The codling moth caterpillar is the infamous worm in the apple.) A few are highly specialized plant feeders. The caterpillars of some tiny moths are leaf miners that spend their larval lives eating the thin layer of soft tissues between the upper and lower epidermal layers of a single leaf. Some make blotch-shaped mines and others make long, serpentine mines that wind about within a leaf, becoming progressively wider as the caterpillar grows. An especially complex caterpillar mine that looks like intricate fretwork is found in aspen leaves. This caterpillar mines from the tip of the leaf toward its base, making numerous sharp, hairpin turns as it moves from the center of the leaf out to and back from the margin in order to avoid the thicker portions of the major veins. The larvae of some wasps, flies, and beetles are also leaf miners.

Some caterpillars are gall makers, as are various other insects

including some species of aphids, flies, and wasps. A gall is a tumorlike swelling of plant tissue that is caused by the insect and that serves as its dwelling place and food. Since the form and structure of the gall and the plant species on which it occurs are usually characteristic of the gall maker, it is often possible to identify the species of gall maker by just looking at its gall. Most gall makers live and feed in a central chamber of the gall, which is lined by nutritious tissue that has a high content of protein, sugar, and fat.

By examining dead goldenrod stems in winter, you can find galls of two types, one caused by the caterpillars of small moths and another made by fly maggots. The gall made by a moth is a spindle-shaped swelling of the stem that is an inch or more in length. In most of these galls you will find a small round hole, the exit from an escape tunnel that was cut by the caterpillar in preparation for its eventual emergence as a moth. The caterpillar must do this because, as you already know, after it metamorphoses to become a moth, it does not have chewing mouthparts and thus can not make its way out of the gall. If you open one of these galls, you will find it to be empty. The moth emerged in the autumn and laid eggs that will not hatch until next spring.

The gall made by a fly is a spherical swelling about an inch in diameter. The escape tunnels in these galls are not easily found. The maggot made an escape tunnel in autumn but left it covered by a paper-thin layer of the gall's outer surface. If you open one of these galls, you will find in the central chamber a full-grown maggot—often used as bait by people who fish through the ice in the winter. The maggots diapause during the winter and emerge through the escape tunnel as adults the following spring. During the winter, they are often devoured by chickadees or downy woodpeckers that peck large holes into the gall.

There are probably less than a hundred species of moths and butterflies that, in the caterpillar stage, eat other insects—only these few out of a worldwide total of about 113,000 species. As you have already seen, caterpillars of the North American harvester, a butterfly of the family Lycaenidae, prey on aphids. Some other lycaenid butterflies also prey on other slow-moving or sessile, permanently attached, insects such as scales. Other insect-eating caterpillars belong to various moth families, among them certain Hawaiian inchworms that belong to a genus with no common name, *Eupithecia*. Species of this genus are the only caterpillars known to have evolved the ability to capture such fast-moving insects as flies, crickets, and parasitic wasps.

Of the fourteen species of *Eupithecia* that have been studied on the Hawaiian Islands by Steven L. Montgomery of the University of Hawaii, only one is a plant feeder; the others are all predators; all are ambushers of insects; and all have similar prey-capturing behaviors. They are well camouflaged so that they can catch unaware the insects on which they prey. Some species are green and lurk motionless and stretched out along the margins of green leaves, while others are various shades of brown or gray and sit motionless on twigs as they assume the same camouflaging, stubby-twig pose that is used by plant-feeding inchworms. When one of these predaceous inchworms, camouflaged and motionless, feels an insect accidently touch its abdomen, it strikes back and grabs the insect with its six front legs, each of which is armed with a specially adapted, long and sharp claw that can pierce the body of the prey. The entire strike takes only about a half second, and the caterpillar then devours its prey. The ambushing tactics of predaceous *Eupithecia* probably evolved from a protective behavior common to plant-feeding inchworms and other caterpillars. They try to drive off predaceous and parasitic insects by striking back with their jaws when their body is touched. A *National Geographic* article by Robert Sisson presents color photographs of a *Eupithecia* larva catching a fly.

Although the adults of cecropia and its close relatives have vestigial mouthparts and do not feed, the great majority of other moths and butterflies have highly specialized mouthparts, a long proboscis suitable for sipping liquids, particularly nectar from flowers. In its resting position, the proboscis, which in some species may be even longer than the rest of the body, is coiled out of harm's way beneath the head. It can be uncoiled to its full length by blood pressure, much like those paper-tube party favors that uncoil when you blow into them. The length of the proboscis varies from species to species. Hawk moths, for example, feed from flowers such as petunias or honeysuckle, in which the nectar is hidden in the bottom of a long flower. They have correspondingly long proboscises. Based on his knowledge of an orchid from Madagascar that has an extraordinarily long nectary, Charles Darwin predicted that a moth with a ten-inch proboscis would eventually be found and shown to be the pollinator of this orchid. A moth with that long a proboscis has since been found on Madagascar, but it has not yet been caught in the act of taking nectar from the orchid in question.

Some butterflies drink things that most of us would not consider to be in character for such dainty sippers of nectar. Many of them have a taste for fermenting substances. I have seen several species, including the beautiful red-spotted purple butterfly, imbibing from fermented fruit lying on the ground. They are sometimes so affected that they almost flop over on their sides. They stagger when they walk, and they lose their usual wariness, often allowing themselves to be picked up with the fingers. I think that these butterflies are stinking drunk. They certainly could not pass a sobriety test.

Moths as well as butterflies seem to find fermenting substances irresistible. One of the most productive and interesting ways to observe or collect moths is to "sugar" for them at night. The "sugar," or bait, is a mixture of canned fruit, molasses, and beer (the alcoholic kind) that has been allowed to ferment until it foams. During the day you paint this mixture on the trunks of trees in a woodland. After dark you patrol the baited trees with a flashlight, just observing the stupefied moths or picking up the abnormally docile creatures to be added to your collection. This is the best way to observe the lovely underwing moths mentioned earlier.

Butterflies of many species can be seen drinking liquid from carrion, feces, or urine stains on the soil, possibly to obtain protein, sodium, or both. Some owlet moths and geometers, the adult forms of inchworms, have gone a step further. They board mammals to suck sweat, skin oils, or lachrymal secretions from the eyes or oozing blood from wounds such as those made by blood-sucking flies.

A southeast Asian owlet moth called *Calyptra,* a member of the same family as our familiar underwing moths, has taken the ultimate step to obtain body fluids. It pierces the skin and sucks blood from mammals such as cattle, pigs, antelopes, tapirs, and even elephants. Under experimental conditions, it will suck blood from humans. This moth's proboscis, originally used for the vegetarian pursuit of sipping nectar, has evolved as an organ specialized for taking blood meals. Its tip is hard, sharp, and set with tiny teeth for rasping and tearing tissue.

Hans Bänziger, the Swiss entomologist who discovered the blood-sucking behavior of these moths, described how he used himself as a guinea pig in his experiments with *Calyptra.* He suspected that this species is a blood-sucker, but no one had ever shown that a moth can pierce the skin of an animal. In his first experiment he made it easy for the moth by first cutting his finger with a scalpel. When he put his hand into a cage containing one of these moths, it climbed onto his finger and plunged its proboscis into the pooling blood. But instead of just drinking the oozing blood, as some other moths were known to do, *Calyptra* "stuck its straight, lance-like proboscis into the wound and . . . penetrated the flesh." Bänziger goes on to say that the pain he felt caused him to utter a cry of joy because he had made an exciting, new discovery. In later experiments he showed that this moth can readily penetrate unbroken skin.

The mouthparts of adult flies are unique among the feeding organs of animals. They make it possible for flies to eat such difficult-to-handle things as dry encrustations of soluble substances such as honeydew or thin films of liquid, perhaps fresh honeydew or the moisture on the surface of dung. The proboscis of one of the familiar flies—a house fly or a blow fly, for example—is thick and stubby and ends in two adjoining spongelike pads. A sucking tube extends from the mouth at the base of the proboscis to the tip of the proboscis, where it ends between the two sponges. When a fly feeds on a thin film of liquid, perhaps

fresh honeydew on a leaf, it presses the pads to the leaf, sponging up the liquid to form a pool between them. The sucking tube then carries the liquid from this pool up into the mouth and on into the digestive system. Flies also use their spongy pads to ingest dry substances such as old honeydew. They press the sponges against the encrustation of honeydew, salivate, and regurgitate onto it, and then sponge up the resulting mess, which will have some of the honeydew dissolved in it.

The maggots of flies have been so greatly modified by evolution that they are barely recognizable as insects. Like the larvae of other insects with complete metamorphosis, they show no externally visible trace of wings. Maggots also lack legs, and their heads, unlike those of any other insects, including the larvae, or wrigglers, of mosquitoes—which are actually primitive flies—have almost disappeared. Most of what little remains of the head has been retracted into the body. The only externally visible parts of the head are two tiny bumps, which may be vestiges of the antennae, and a pair of small, sharp appendages known as mouth hooks, the only remaining functional parts of the greatly simplified organs of feeding. The maggot uses the mouth hooks for various purposes, including the tearing and rasping of solid food into a slurry that it can swallow.

Adult hover flies, well known as mimics of bees or wasps, are commonly seen taking nectar from flowers, sponging honeydew from leaves, or drinking sap oozing from trees, especially from maples in the early spring. Although some of them may eat pollen or sip juices from excrement or carrion, most of them depend solely on plant exudates for sustenance. The feeding habits of the larvae are more varied. Some eat and live in rotting wood; a few are scavengers in the nests of bumble bees or ants; a few feed in the living bulbs of plants such as narcissus; and others live in carrion or sewage. Among the last category is the drone fly, a convincing mimic of the honey bee, which spread from its original home in Eurasia to the rest of the world as sewage-disposal plants proliferated.

The larvae of some of the more commonly seen hover flies prey on aphids. The harmless adults of these aphid-eating species are barred with yellow and black and more or less resemble bees or wasps. Although they cannot sting or bite, they sometimes alarm people by hovering near the face or landing on a bare arm or leg to sponge up a little sweat. They visit flowers for nectar and are often seen feeding on honeydew.

The larvae, having hatched from an egg that the mother placed nearby, live in the midst of an aphid colony, surrounded by their almost helpless prey. They are often accompanied by other aphid-eating insects such as adult or larval ladybird beetles. The aphids sometimes seem to pay little attention to these predators, continuing to drink plant sap as their nearest companions are being devoured. As Harold Oldroyd says in his fascinating book *The Natural History of Flies*, the hover fly larva finds its prey by swinging the front end of its body from side to side until it touches an aphid. It then pierces the victim with its mouth hooks and sucks it dry as it holds it high in the air where it cannot crawl away. A hover fly larva may eat as many as fifty or sixty aphids in a day. It is no wonder that farmers and gardeners consider them to be beneficial. Some aphids, however, produce an alarm pheromone when predators or other enemies are present. In response to this pheromone, the aphids in a colony may space themselves more widely from each other or may even allow themselves to fall from the leaf on which they are feeding.

Black blow flies drink nectar when they are adults, but as maggots they eat the flesh of dead animals, thus performing an essential ecological function. Together with related blow flies and other scavengers, ranging in size from bacteria to vultures, they are the all-important recyclers in many of the world's ecosystems. They return dead plants and animals to the soil, whence their atoms and molecules are resurrected to form the bodies of future generations of plants and animals.

You may be amazed to learn that carrion-eating maggots, including black blow flies, play a small but important role in the practice of medicine. Improbable as it may seem, they are used to disinfect wounds. When surgery and antibiotics fail, they are the last resort for eliminating osteomyelitis and other infections that are deep in the tissues of the body. Once you understand how these maggots behave, using them to cleanse wounds does not seem improbable at all. Unlike screwworm flies, they are carrion feeders and eat only dead tissue. Thus if they are in a wound on a living animal—as occasionally occurs—they eat dead, putrefying tissue but not living tissue. They also secrete an antibiotic substance called allantoin, possibly to kill the bacteria that are their chief competitors for dead tissue. When the maggots are fully grown, they leave the wound of their own accord, dropping to the ground and burrowing into the soil in preparation for the molt to the pupal stage. When used to heal wounds, the full-grown maggots also leave the wound but gather under the dressing that covers the wound, whence they are easily removed.

The knowledge of "maggot therapy" probably goes back to prehistoric times. Using maggots to clean wounds was apparently an ancient tradition of a tribe of Australian aborigines. But the recent history of "maggot therapy" goes back to the days of trench warfare during World War I. Army physicians noted that men whose battle wounds were infested with maggots seldom developed bacterial infections. These men had lain in the no-man's land between the friendly and the enemy trenches for a long time. Wounded men who had been rescued quickly and rushed to an aid station were not infested with maggots but very often developed infections. Physicians, including W. S. Baer, soon put two and two together and adopted the use of specially raised, sterile maggots to treat refractory infections. Maggots were frequently used by physicians for over twenty years, but this practice was all but abandoned after the first antibiotics were discovered.

But to this day some physicians turn to maggot therapy when all else fails. On several occasions during recent years, the Department of Entomology at the University of Illinois has supplied sterile maggots to local physicians. The November 24, 1990, issue of the *Champaign News-Gazette* of Champaign, Illinois, reported that a surgeon, Dr. Adolf Lo of the Covenant Medical Center in Urbana, used sterile black blow fly maggots to cure an infection in a large, deep wound in the leg of a diabetic woman. Surgery and antibiotics had failed to heal the wound. If it had not been for maggot therapy, the leg would have been amputated. As Dr. Lo's nurse, Carol Brenner, said, "When all conventional methods fail, you do what you have to do."

When I was a beginning graduate student and knew less about fleas than I do now, my wife and I lived in an apartment with wall-to-wall carpet in the living room. Our dog slept on this carpet. One summer, we left the dog in a kennel and went off on a trip. We returned three weeks later. My wife, bare-legged and in shorts, walked into the living room as I struggled with luggage on the door step. Moments later she screeched and then shouted, "Why does the wife of an entomologist have to put up with this?" Scores of fleas were jumping on the carpet, and some of them were biting her legs. It surprised both of us. We had never before seen a flea in our apartment.

You must understand the habits of fleas to understand how our domestic crisis came about. Most adult fleas suck blood from mammals, although a few species prefer birds. Like most external parasites that travel through feathers or fur, fleas are wingless. Nevertheless, they are quite mobile by virtue of their powerful jumping legs. They leap on and off of their host, to board it in the first place or to get off to lay their eggs in its bedding. The record jump made by an oriental rat flea is twelve inches, 150 times the length of its own body. If human athletes could perform comparably, the record for the standing broad jump would be at least 850 feet.

The legless larvae live in the litter in the nest, burrow, or other sleeping place of the host. They eat organic debris, including the feces of adult fleas, which fall from the host's fur and consist mainly of the host's blood. The larvae spin a silken cocoon about themselves before molting to the pupal stage. Some species, including those that parasitize cats and dogs, may remain dormant in the cocoon for days or even months after they have become adults, not emerging from the cocoon until they are stimulated by vibrations caused by movements of the host, a sign that a blood meal is available. Gottfried Fraenkel, in a course that I took *after* my wife's experience with fleas, demonstrated this phenomenon

to our class. He gently placed a dish containing dormant oriental rat fleas on a table. When he pounded on the table with his fist, the fleas almost instantly popped out of their cocoons.

The events that led to our flea crisis can now be deduced. As long as the dog was around, we were never bitten by fleas and never saw them. They prefer dogs to humans and were always out of sight in the fur of the dog or the pile of the carpet. While we were gone, the tiny eggs and larvae in the carpet grew to the adult stage but remained dormant in their cocoons because they were not stimulated by people or dogs tramping around. When my wife walked onto the carpet, all of these dormant fleas became active. Ravenously hungry, they jumped onto the only available source of blood, my unfortunate wife. I should have sent the dog in first!

Dog and cat fleas are annoying, but some fleas, notably the oriental rat flea, have had a calamitous effect on human history. By means of their bites they transmit from rats to humans the bacterium that causes bubonic plague, the black death. Known since biblical times, plague has repeatedly raged around the world in great epidemics and pandemics.

From 1348 to 1382, four pandemics swept across Europe, killing 25 million people, about one-quarter of the population of the continent. Moral, religious, and political upheaval followed in the wake of the disease, eventually resulting in the disintegration of the feudal system. A pandemic that began in Turkey in 1661 spread westward across Europe, reaching London in 1665 and ultimately killing 70,000 of its 470,000 inhabitants. The most recent pandemic began in northern China in the late nineteenth century and spread half way around the world, appearing in San Francisco in 1900. Since then plague has smoldered. There have been minor outbreaks and scattered cases around the world. Recently there was an outbreak of plague in India. The plague bacterium persists in wild rodents in the western United States. Almost every year, a few hikers and campers, especially those who sleep on the ground near animal burrows, are infected by fleas from prairie dogs, ground squirrels, or other rodents.

In seventeenth-century Europe, people lived close to rats and took flea bites for granted. We can glimpse the horror of the plague, in a time when its cause was unknown and there were no remedies, by following the progress of the London epidemic in *The Diary of Samuel Pepys*. On September 24, 1664, he first mentioned the plague: "We were told today of a Dutch ship of 3 or 400 tons, where all the men were dead of the plague, and the ship cast ashore at Gottenburgh." In 1665 he wrote:

May 24. All the news is . . . of the plague growing upon us in this towne; and of the remedies against it; some one thing and some another.

June 7. This day . . . I did in Drury Lane see two or three houses marked with a red cross upon the doors, and "Lord have mercy upon us" writ there.

June 15. The towne grows very sickly, and people to be afeared of it; there dying this last week of the plague 112, from 43 the week before.

June 21. . . . I find all the towne almost going out of towne; the coaches and waggons being all full of people going into the country.

August 31. In the city died this week 7,496, and of them 6,102 of the plague. But it is feared that the true number of the dead this week is near 10,000."

Coping with the Seasons

By late August the look of the land has changed. In the few remnants of tall grass prairie left in the midwest, the small annual grasses have turned brown, and perennial species are green and blossoming, such as big bluestem, a giant at six to eight feet in height. Soon its seeds will be eaten and scattered by birds, but some will survive to germinate next spring. Along country road-sides everywhere, goldenrods and thistles are flowering, a gay medley of yellow and pink blossoms attended by the bumble bees that pollinate them. Colonies of these bees are now at their peak and are just beginning to produce males and queens. As you already know, workers and males will all be dead by mid-autumn, survived only by mated queens that will found new colonies next spring.

Just as goldenrods, thistles, and cultivated chrysanthemums are stimulated to blossom and produce seeds in late summer and fall when the days are becoming short, American goldfinches respond to day length and reproduce at this time of year, too. Throughout the summer, when other birds were nesting, flocks of goldfinches wandered woodlands and fields in search of seeds. Now, in late summer, the flocks have broken up and breeding pairs have moved to shrubby pastures and meadows, where they have built nests lined with thistledown in the forks of small trees and are harvesting the now abundant seeds, especially those of thistles, that will nourish them and their nestlings.

Although it is still hot and winter seems far away, many creatures have begun their preparations for the cold season. Monarch butterflies have started their southward migration and sometimes dozens of them can be seen clinging to a patch of goldenrods or thistles as they sip the nectar that will fuel their flight. Birds are also on the move. A few land birds, including some warblers and flycatchers, have begun their slow journey to Central and South America. Blackbirds have formed flocks, and these mixed groups of red-winged blackbirds, brown-headed cowbirds, and common grackles will roam the countryside until they head south in the late fall. Shorebirds such as dowitchers, godwits, plovers, and sandpipers began to leave their Arctic nesting grounds in June. Their southward migration through our area was already well under way in July, and is at its peak now. Many insects have already committed themselves to spending the winter in the hibernation-like state of diapause. Most of the rest of them will make this commitment within the next few weeks.

In temperate climates, each of the seasons presents its own threats to the survival of insects. The heat and dryness of midsummer are obstacles to some species, but the cold and near-starvation conditions of winter are serious threats to the survival of virtually all temperate zone insects. Only a handful of them can remain active during the winter, and even fewer of them can find food at that time of year. All of the rest have found other ways to survive the inhospitable conditions of winter. A few of them migrate to warmer climates, but the great majority of temperate zone insects survive by entering diapause and staying put as they brave the winter.

Among the very few insects that are active during temperate zone winters are snowfleas, snow scorpionflies, winter crane flies, and winter stoneflies. Some of these insects may be in a diapause that permits activity but blocks reproduction, but that has not been clearly established. As have most of the other non-tropical insects, these species have evolved ways of keeping their bodies from freezing. But they have gone beyond that. They have also evolved ways to remain active at very low temperatures that would leave other insects torpid and incapable of movement. On mild days, adult winter stoneflies can be seen clinging to rocks, tree trunks, or the abutments of concrete bridges along the unpolluted streams in which their nymphs live. Snow scorpionflies, relatives of the kleptoparasitic scorpionfly that steals prey from the webs of spiders, eat the moss under the snow but also crawl about on its surface. Winter crane flies, true flies with vestigial wings too tiny for flying, are sometimes seen crawling on snow, and the so-called snowfleas, not really fleas at all but tiny, dark-colored springtails, may gather in large groups as they hop about on the surface of the snow.

The ability of these and other insects to remain active during winter enables them to exploit habitats at a time when their other inhabitants, both predators and competitors, are quiescent. Food is not necessarily a problem for all insects that are active in the winter. Some of them survive on fat stored during a time of greater abundance. Nevertheless, minimal food resources are available in winter. It is a time of starvation for insects that eat other insects or the leaves of deciduous plants, but some foods are present. Green algae survive on rocks and the bark of trees. The moss is still green under the snow, and even the surface of the snow becomes scattered with wind-blown pollen and bits of organic debris.

The rock crawlers, more often known by the scientific name of *Grylloblatta*, live in perpetual winter conditions. These thoroughly cold-adapted insects, which seem to resemble both crickets and cockroaches (*Grylla* means cricket and *blatta* means cockroach), occur only high in the mountains of Japan, Siberia, and northwestern North America. They live at the edges of glaciers and snow fields and can be found crawling about in moss, in ice caves, or on the surface of snow, presumably feeding on dead insects and other organic matter, much of it probably blown up from lower elevations. So completely are they adapted to the cold that they are killed by the heat at temperatures that would still be too cold for most other insects. Rock crawlers are active between a low temperature of only 27° F and a high temperature of a chilly 53° F. They cannot survive temperatures that are much higher. At 64° F they become torpid and at 69° F they are irreversibly damaged by the heat and ultimately die. While the preferred temperature of field crickets is about 88° F, typical of a warm summer day, the preferred temperature of *Grylloblatta* is just slightly above freezing, about 35° F.

Unlike rock crawlers, most insects survive the winter by entering diapause and remaining inactive. In a diapausing insect, physical development to the next life stage stops. Thus a diapausing egg does not hatch or resume embryonic development until the state of diapause has been terminated; nymphs or larvae and pupae do not metamorphose to the next stage; and diapausing adult females do not lay eggs or give birth to young. Most diapausing insects are almost totally inactive, but a few of them, especially adults, may be more or less active and may even drink water or eat a little from time to time. As you will see later, the monarch butterfly is an extreme example of an insect that remains active while it is in diapause.

Insects in winter diapause usually have physiological mechanisms that allow their bodies to supercool: their body temperature can fall well below the freezing point without the formation of ice crystals that would irreparably damage the cells in their body and thus cause them to die. Insects generally achieve protection against freezing by producing antifreeze. Much as we add alcohol or some other substance to our car radiators in winter, insects permeate their bodies with internally secreted glycerol or sometimes some other alcohol, such as sorbitol or mannitol. Finally, just as you would suspect, the metabolic rates of diapausing insects are far lower than those of nondiapausing insects—usually from one-tenth to one-fifteenth of the usual rate, and sometimes only one-twentieth or even less. An insect whose metabolic rate falls to one-tenth of the usual reduces its consumption of energy by 90 percent and can thus survive on its stored body fat for about ten times as long as can a nondiapausing insect. Hence, most diapausing insects can survive the winter without eating.

To diapause or not to diapause? That is a decision that virtually all temperate zone insects, and some tropical insects too, must make at some time during their lives. Would it be better to continue developing now—to mature, produce offspring, and then die, or would it be better to go into diapause and thus defer reproduction until a more favorable season?

The critical factor in deciding when to diapause is whether or not there will be enough time to complete another generation before an unfavorable period of winter cold or summer heat or dryness sets in. Such unfavorable conditions will probably kill the young of the next generation before they can grow to the life stage in which they themselves can enter diapause and thus be protected against the rigors of the heat or the cold. The exact timing of the initiation of diapause varies from species to species. It depends largely on the length of time required for a generation to mature. For fast-growing insects such as flies, the span of a generation may be only a week or two, and many generations can be completed in a summer. Thus these insects need not make the decision to diapause until the onset of winter is relatively close. But slow-growing insects such as cecropias have only one generation per year, the completion of a generation requiring more than half of the summer. In this case, the decision to diapause must be made well in advance of the beginning of the cold season, when the insect is still a partly grown caterpillar.

Since diapausing insects stop developing and lower their metabolic rate, they are essentially in a state of "suspended animation," and can survive a long northern winter, the hottest part of a summer, or a dry season in the tropics as they rest hidden away in some protected nook. Insects as a group can diapause in any one of the life stages, but members of the same species generally diapause in the same life stage. Eastern tent caterpillars, walking-sticks, and most aphids diapause as eggs; dragonflies and cicadas as nymphs; codling moths and fritillary butterflies as larvae; cecropias and tiger swallowtails as pupae; and squash bugs, chinch bugs, and mourning cloak butterflies as adults.

An insect requires a cue to tell it when the proper time for its species to go into diapause has arrived. The cue that warns of approaching winter must be unfailingly reliable. If diapause occurs too early because of an unreliable cue, the opportunity to produce another generation will be lost. If diapause occurs too late, the next generation will not survive the winter with its lethal cold and lack of food, such as green foliage or other insects. Temperature alone is not a reliable cue. For a slow-growing insect the appropriate time to enter winter diapause may be in August, long before the first cold of autumn. Even if the appropriate time were in late September or October, some years have unusually warm autumns, and cool weather may be delayed until November. Thus the insect that uses the onset of cold weather as a cue to the approach of winter might not enter diapause early enough and might produce a second generation that would be doomed to die when cold weather does arrive.

The only completely reliable cue is the length of the day. In the North Temperate Zone, December 21, the first day of winter, is the shortest day of the year. On that day in central Illinois there are fewer than nine and a half hours of daylight plus about half an hour of twilight. June 21, the first day of summer, is the longest day of the year, slightly over fifteen hours of daylight plus another one-half hour of twilight. From June until December, each succeeding day is a few minutes shorter than the preceding one, auguring the approach of winter. From December until June, each succeeding day is a few minutes longer than the preceding one, foretelling the arrival of summer. Thus each day of the summer/fall season and each day of the winter/spring season has its own typical length. For example, there are only two days in the year when there are exactly twelve hours between sunrise and sunset, the first day of autumn, usually about September 21, and the first day of spring, usually about March 21. Some insects are triggered to enter summer diapause by the relatively long days of late spring and early summer. But most insects enter winter diapause in response to the relatively short days of late summer and early fall.

By late summer, cecropia caterpillars have, in response to the short days, committed themselves to diapause. They show no outward sign of having made this commitment. Indeed, they will continue to feed until they grow to full size. Then they will spin their cocoons and molt to the pupal stage within the confines of the cocoon. The pupa, like the caterpillar, shows no outward sign of being in diapause, but it will remain dormant until the following spring, when development to the adult stage will resume.

Under natural conditions anywhere in the United States and Canada, the days of late summer are sufficiently short to trigger diapause in cecropia larvae. But we can prevent the onset of diapause by keeping these caterpillars in the laboratory and creating artificially long days with electric lights. If we do this, the insects will begin metamorphosing to the adult stage immediately after they molt to the pupal stage. When the adults emerge about three weeks later, they will lay eggs and we could, if we wished to do so, raise extra generations in the laboratory by feeding the caterpillars an artificial diet composed of casein, sugar, vitamins, and other nutrients in an agar gel.

In the spring cecropia pupae and other temperate zone insects terminate the state of diapause in order to resume development and reproduction. How do they know that warm days are ahead and that it is safe to resume normal life? The first warm days of spring do foretell the approach of summer, but if an insect were programmed to terminate diapause in response to a few days of warmth, the warmth of autumn or a spell of warm weather in winter could trigger the premature termination of diapause and thus doom the insect to death when the usual cold temperatures of winter return.

How can this dilemma be resolved? How can insects cope with the variability and unpredictability of the weather? Cecropia and many other insects avoid the deadly trap of premature diapause termination because they are programmed to respond to a period of warm weather only *after* they have been exposed to a sufficient duration of cold weather. Cecropia keeps tabs on the amount of *coldness* that it has experienced—a process that is akin to the meteorologist calculating and keeping a total of heating degree days in winter. Under natural conditions, a cecropia pupa will terminate diapause in response to a period of warmth only after it has been sufficiently chilled.

If you can find a few cecropia cocoons in autumn, you can demonstrate to yourself the effect of chilling on diapausing pupae. First, check the cocoons to make sure that they contain living pupae. Gently shake each one next to your ear. If the cocoon is heavy and you hear a solid thud, it probably contains a live pupa. If it is light and rattles, it probably contains only a dead and dry caterpillar. Then put a few of the good cocoons in a warm place in your home, and put an equal number of them in the refrigerator. After ten weeks remove the cocoons from the refrigerator and place them in the warm place beside the other cocoons. After about three weeks, moths will begin to emerge from the chilled cocoons, and most of them will emerge as a relatively synchronized group within two or three weeks. The unchilled pupae will behave differently. Most likely none of them will have emerged by the time you take the chilled cocoons out of the refrigerator. One or two moths may emerge from unchilled cocoons after six or seven months; another few may emerge after a year or so; and the rest of them will probably die without ever metamorphosing to the adult stage.

Cecropias, and other insects as well, have also evolved a way to cope with other vicissitudes of the weather: the rare frost near the end of May that kills many insects and dooms the Illinois peach crop, the occasional deadly drought of midsummer, or the once in a hundred years phenomenon of freezing temperatures in late summer; at Freeport, Illinois, the average date of the first frost in autumn is October 7, but in one year there was a killing frost there on August 30. One way that an insect can deal with such rarely occurring contingencies is to divide its progeny into two or more groups that emerge at different times of the year. Cecropia moths, for example, produce progeny of two types: one group that is destined to emerge early the next spring because the insects terminate diapause as early as possible after they have been sufficiently chilled, and another group that is destined to emerge in late spring because the insects terminate diapause much later than does the first group.

The moth from the dense shrub below the birch that you read about in the first chapter is just one member of the early-emerging group, which generally includes only about 10 percent of the cecropia moths that will emerge from their cocoons in a given spring. When these early moths are on the wing during the last two weeks of May, the other 90 percent of the overwintering cecropia population has not even begun to metamorphose to the adult form. It is not until the first few days of June, about two months after the early pupae began to metamorphose, that the members of this late group show the first signs of the development of the adult within the pupal skin. Because temperatures are higher in June than they are during April and May when the early emergers develop, these late emergers require only a little more than two weeks to complete development, rather than the nearly two months required by the early moths. They emerge during the last two weeks of June and the first week of July, just after the summer solstice, and then they follow the same pattern of behavior as the early-emerging moths.

Ecologists who have considered organisms with emergence patterns like cecropia's are generally agreed that two-part emergences are advantageous because they are bet-hedging strategies. Whether or not an individual appears as a moth early or late is genetically determined, and any pair of parents produces some offspring that ultimately emerge with the early group and others that ultimately emerge with the late group. They thus hedge their bets by not putting all of their money on one seasonal horse. In a year with an early spring, the first group will do well, and may be the only survivors if their late-emerging brothers and sisters are killed by a severe summer drought or an August freeze. Conversely, the members of the early group, or even their eggs or young larvae, may be killed by a late spring freeze. In this case, the late group might survive to assure the immortality of their parents' genes.

Our first inkling of cecropia's two-part emergence came when my friend and colleague James G. Sternburg used his screened porch as a trap for wild cecropia males. Each night from early May to mid-July, Jim baited his porch with a virgin female in a small cage that prevented her from mating with any males that might appear, thus assuring that she would continue to release her sex-attractant pheromone after the arrival of the first male, as well as on succeeding nights. The outer screen door of the the porch was left ajar. The males entered easily but seldom found the way out. Data from the trapped males were, of course, indicative of the emergence pattern of both sexes. Males and females emerge from their cocoons at approximately the same time, although males tend to emerge two or three days ahead of females. I soon became fascinated by Jim's project and he graciously suggested that I join in. Thus began a long partnership in which we found great pleasure in solving some of the riddles of cecropia's existence.

In case you are wondering, the bait females came from cocoons that Jim stored in a refrigerator. We induced them to emerge when we needed them by moving them to a warm place about three weeks in advance.

Jim's results surprised us. We expected one continuous emergence, the usual pattern with most insects. But there were instead the two distinct groups that I have already described. A few males appeared in late May, then there was a period during which no males were caught, and then a much larger group appeared in late June. We later discovered that in 1909 Gene Stratton Porter, author of the classic *A Girl of the Limberlost*, wrote in her *Moths of the Limberlost* that cecropia moths appear "in middle May or June" (my italics). Her discovery preceded Jim's by almost fifty years, but her words went unnoticed by entomologists or other biologists.

The next year we did our first controlled experiment, mainly to see if the two-part emergence would happen again but also to find out if the date on which a moth emerges is determined by the microhabitat in which the pupa spends the winter. Some cocoons are exposed well above ground on the twigs of trees, but many more are spun close to ground level among the stems of a shrub or the adventitious suckers at the base of a tree. By late autumn, the low cocoons are blanketed with fallen leaves. We thought that the high cocoons, more exposed to the sun, might account for the small early-emerging group and the more abundant low cocoons for the larger late-emerging group. As it turned out, we were wrong. It is not a matter of where the cocoons are located. Genetic studies that we did later showed that each individual has already been destined to emerge with one group or the other long before it spins its cocoon—even before it begins embryonic development in the egg.

We began our experiment by collecting cocoons during winter from shade trees and shrubs along the residential streets of the twin cities of Champaign and Urbana, Illinois. The idea was to put these cocoons in a screened enclosure where they would be exposed to the weather. Although the weather varied, all of the cocoons in the enclosure were exposed to the *same* conditions. Thus if early and late emergence are determined by *differences* in the pupa's exposure to the elements in the microhabitat in which its cocoon is located, the moths in our enclosure should not have emerged from their cocoons in two distinct groups. But they did.

Since we had collected only along the streets, a population of cocoons still remained in the backyards of the residential areas. These were the control group that revealed the emergence pattern from cocoons that were left undisturbed in their various microhabitats. We sampled this group by catching males in large traps that were baited with female cecropias from the beginning of May to the middle of July. One trap was in my backyard in Champaign, and the other was five miles to the east in Jim's backyard in adjoining Urbana. In keeping with our emergence data, we caught a small number of males in late May and a much larger group in late June. One morning in June, I found 240 males in my trap. That many huge moths—the wing span of the male is about five inches—was overwhelming! The moths from the traps and those that emerged in our enclosure corresponded almost perfectly in the relative size of the two groups and in emergence date, although the trap catch did peak two or three days later than did the emergence group. This was to be expected since the bait females attracted more and more males as the number of competing wild, virgin, pheromone-releasing females steadily declined.

Although cecropia is not alone in having a two-part emergence pattern, most adult insects appear in the spring as a single coherent and usually well-synchronized group. This pattern has its advantages, especially for insects that may live for only a few days, such as cecropia. Emerging with the crowd enhances an individual's chance of finding a mate. Furthermore, a well-synchronized emergence usually ensures that all individuals coincide with resources that may be available only briefly. For example, some hover flies that are choosy about which blossoms they visit for nectar are present as adults for only about three weeks in the spring, the period during which the chosen plants are in flower. Why, then, does cecropia divide its forces and emerge in two groups that are separated by a month and do not overlap with each other? The answer is, of course, that cecropia has evolved the bet-hedging strategy that I have already described. If you will permit a switch in metaphors, it does not put all of its eggs in one seasonal basket.

Some insects hedge their bets to an even greater extent by making a small side bet on a long shot: that their offspring may be more likely to survive next year than this year. A very few cecropias, for example, do not emerge from their cocoons until they have passed through a second winter. If none of a pair's progeny—of either emergence group—survives in an unusually unfavorable year, the next year might be better, and the few individuals that emerge in the next year might live to pass their parents' genes on to another generation.

The value of this extended bet hedging to the insect is demonstrated by the research that James L. Krysan and his colleagues did with one of our native beetles, aptly known as the northern corn rootworm. In prehistoric times, the larvae of this beetle fed on the roots of tall prairie grasses, but now they eat the roots of another species of grass: corn, or maize, one of our most valued crops. This insect may have attacked maize grown by native Americans, but it definitely became a pest when the prairies were plowed for farming by European settlers. In late summer the beetles lay their eggs in the soil around living corn plants. The eggs are in diapause and do not hatch until late the following spring. If the farmer plants corn in that field for a second season, the larvae prosper by eating its roots. But if the farmer plants something else, the tiny, wingless larvae die because they will eat only corn roots and because they are too small to crawl to another field. For decades many midwestern farmers avoided the depredations of this insect by crop rotation—and many still do—by

not growing corn in the same field for two years in a row, planting soybeans instead in alternate years.

But because of their extended bet hedging, northern corn rootworms can now survive crop rotation in some parts of the Corn Belt. A very few eggs always passed through two winters before they hatched. With crop rotation, those that hatched in the first year died because no corn grew in the field. The few that hatched after the second winter survived because corn was once again planted in the field. Over the decades, beetles with the gene for delayed emergence supplanted the others, and consequently, in some areas of the midwest, crop rotation no longer protects the corn crop.

The supplanting of northern corn rootworms that diapause for only one winter by northern corn rootworms that diapause for two winters is another example of natural selection in action. It is a process that began within living memory and that continues now. Technically speaking, this example of selection is not really "natural," because the selective force is created by humans, farmers who rotate their crops and thus cause the rootworms' food to disappear in alternate years.

Migration is another way of escaping the rigors of winter. We all know that many of the birds that nest in temperate areas—especially the insectivorous species—are migrants. In autumn they and their northern-born progeny move south to subtropical or tropical climates, where food is available and winters are mild or nonexistent. The following spring they return to the north to nest and raise their young, taking advantage of the bounteous food supply resulting from the seasonally concentrated and exuberant flush of summer growth. Similar migrations are made by quite a few insects, the monarch butterfly notable among them. In a few insect species, the same individuals make both the southward and northward journeys, as do birds. But with insects the usual situation is that the children or grandchildren of the migrants make the return trip. Sometimes migrants stop to breed along the way so that the southward or northward trip, or both, is completed by the leapfrogging of succeeding generations.

The potato leafhopper breeds in the states along the Gulf of Mexico throughout the year, but is not found to the north of these states in the winter. Early each spring, these tiny, green insects—an adult is barely one-eighth of an inch long—fly up into the atmosphere to be carried as far north as Canada by warm southerly winds. Although the wind carries these insects, they must beat their wings in order to remain at the altitude at which the wind blows. They pass through several generations in the north and often become so abundant that they are important pests of potatoes and other crops. In southern Wisconsin I once saw them so numerous that a farmer driving a tractor through a field of lima beans was surrounded by a cloud of potato leafhoppers so dense that he had to wear a respirator. Some entomologists have thought that these leafhoppers do not make a return migration to the south, that northern populations die off each winter and are replaced by more migrants from the south each spring. I have my doubts about this, although there is as yet little evidence of a southward migration. After all, no matter how successfully potato leafhoppers reproduce in the north, there can be no natural selection for migration if it is a one-way dead end that ultimately results in the death of the northward migrants and all of their progeny.

Recently, a team led by Joseph R. Riley of the Radar Entomology Unit of the United Kingdom's Natural Resources Institute showed that the tiny brown planthopper, which belongs to a family closely allied to the leafhoppers, does make a return southward migration in autumn after its northward migration in spring. This insect, one of the most serious pests of rice, the world's major grain crop, can survive the winter only in tropical and subtropical regions of eastern Asia. Its northward migration is much like that of the potato leafhopper; southerly winds carry spring migrants to temperate rice-growing areas of China, Korea, and Japan. Riley and his colleagues provided strong evidence that the descendants of spring migrant brown planthoppers do make a return migration in autumn. Riley's radar, specially designed to detect insects, showed that on late September evenings there are dense, southward-moving swarms of tiny insects at altitudes of from 1,300 to

3,300 feet. Traps sent into the swarms on balloons showed that the great majority of these insects were brown planthoppers.

Hugh Dingle of the Department of Entomology at the University of California at Davis studies the migration of the milkweed bug, which, like the monarch, stores in its body toxins from the milkweed plants on which it feeds. While monarch caterpillars chew the leaves of milkweeds, milkweed bug nymphs and adults both suck juice from the seed pods. (Like the monarch, milkweed bugs are patterned in orange and black to warn off predators. In fact, most of the insects that feed on milkweeds are orange and black or red and black.) These bugs occur from Central America to Canada, but survive the winter only in the southern parts of their range. After arriving in the north in the spring and summer, they undergo rapid population growth, and it is their descendants that make the return migration, probably by leapfrogging south from generation to generation.

Milkweed bugs are not abundant enough to track with radar, and neither are they big enough nor conspicuous enough to follow by means of mark and recapture techniques, which, as we will see, have been used to track monarchs. Dingle used a different technique, taking advantage of the tarsal reflex that seems to occur in all winged insects. If the tarsi (feet) of an insect lose contact with the surface, the insect will by reflex begin to fly and will continue to do so for some time, especially if a gentle breeze is blown in its face. You can try this yourself with a fly or some other insect with a string or a thin stick glued to its back just in front of the wings. Let the insect crawl on a table and then pull it up into the air. Usually, it will begin to fly on the first try. Let the feet touch the table and it will stop flying. Hand the insect a small wad of cotton or ball of Styrofoam, and it will grasp it in its feet and also stop flying. Take this object away, and the insect will fly again.

Dingle glued a thin stick to the milkweed bug's upper side just behind the head and perpendicular to the long axis of its body, so as not to interfere with the motion of the wings. Sure enough, the insect flew when he raised it up from the surface on which it sat. He could now clock the duration of these tethered flights as a measure of the tendency to migrate. Bugs that were tested in tethered flight in the summer made short flights of a few minutes' duration, but those that were tested during the shorter days of early fall generally made long flights that often continued for two or three hours. The obvious conclusion is that the bugs that make short flights in the early summer are not yet ready to migrate, but those that make long flights in the fall are ready to migrate south. The occurrence of long flights in the laboratory corresponds to the autumn disappearance of the bugs from their natural habitats. Dingle found that short days also delay the onset of egg laying by well over a month—until migrating bugs have arrived in a warm area to the south. Thus the migratory bugs are in a reproductive diapause, as are many migratory adult insects, including the monarch. Unlike most insects in reproductive diapause, monarchs, milkweed bugs, and some others are not protected against cold weather, but they can carry on such nonreproductive activities as feeding and flying.

On a warm, sunny day in August, a monarch sips nectar from ironweeds that grow near her birthplace a few miles north of Sault Ste. Marie, Ontario. The northernmost limit of her species is less than a hundred miles away. When she finishes her sugar-rich meal, she begins to fly south at a cruising speed of about ten miles per hour. Like the milkweed bug, she is in reproductive diapause. A little later that morning she crosses the St. Mary's River and the Soo Canal to enter the Upper Peninsula of Michigan. During the day she stops to feed at flowers. At night she rests in the foliage of a tree or bush, often in a tight cluster with a dozen or more other monarchs. Less that two weeks later, she flies through central Illinois, stopping to feed at blossoming goldenrods. She is on a journey of almost 2,000 miles that will take her from her northern birthplace to her wintering site high in the Sierra Madre Mountains just west of Mexico City. Although she herself has never made the trip before, she will unerringly find her way to a small group of pines, only a few acres in extent, on which her ancestors spent the winter for many years, probably for many centuries, in the past.

In this patch of pines, as in eleven other similar overwintering sites scattered through the nearby Sierra Madre, monarchs gather

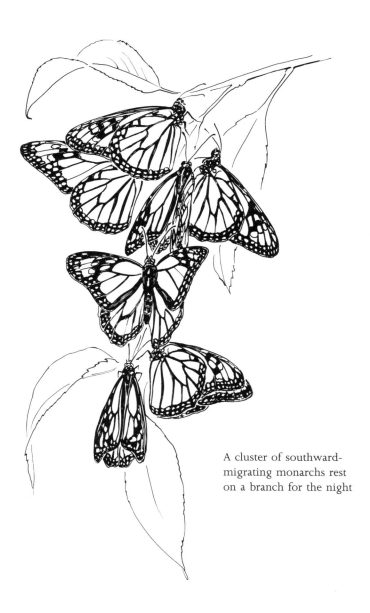

A cluster of southward-migrating monarchs rest on a branch for the night

by the tens of millions. It has been estimated that one 5.5-acre site contained about 22.5 million monarchs, about 4 million butterflies per acre. So dense was the concentration that the trees appeared to be orange, the clusters of butterflies obscuring the green needles of virtually every twig and branch. At the wintering sites, all at about 9,000 feet in elevation, it is often cold and there may even be frosts. The monarchs are all but torpid and spend most of their time clustered in the trees. But on warm days they fly to nearby areas to drink water or nectar.

In March, about four and a half months after they began to arrive at the overwintering site, their reproductive diapause has ended and they begin to court and mate. Shortly thereafter, they move northward, and they or their descendants are back on their breeding grounds by mid to late spring. As Steve Malcolm of Western Michigan University in Kalamazoo recently told me, some returnees seen as far north as Oklahoma are badly worn and probably made the journey all the way from Mexico. Others are fresh and bright, apparently the offspring of migrants that interrupted their northward journey to breed along the way.

The first of the overwintering sites of the monarch in Mexico was discovered in January of 1975. But the search began in 1937 when the Canadian zoologist Fred A. Urquhart and his wife, Norah, began gluing paper tags to monarchs, each reading "Send to Zoology University Toronto Canada." Thousands of tagged monarchs were released in the hope that they would be captured and returned from points along their southward route. Over the years, thousands of people all over Canada and the United States cooperated with the Urquharts in tagging and capturing monarchs. Many marked butterflies were returned to the University of Toronto, and eventually a migration pattern became apparent. The monarchs migrated from northeast to southwest. They funneled into Texas and a few were recaptured in Mexico. But no overwintering site was found until Cathy and Ken Brugger of Mexico City, two of the Urquharts' collaborators, found the first of the Mexican sites in January of 1975.

Urquhart and his wife went to Mexico to see the site. As he said in an article in a 1978 *National Geographic* magazine, he and his wife were no longer young and the climb to the site was difficult for them. But it was worth it. They were met by the awesome sight of millions of monarchs hanging from the trees and littering the ground. Urquhart topped off his day by making an amazing find. He reported the incident in the *National Geographic*:

> While we stared in wonder, a pine branch three inches thick broke under its burden of languid butterflies and crashed to earth . . . I stopped to examine the mass of dislodged monarchs. There, to my amazement, was one bearing a white tag!
>
> By incredible chance I had stumbled on a butterfly tagged by one Jim Gilbert, far away in Chaska, Minnesota. Later Mr. Gilbert sent me a photograph of the very field of goldenrod where he had marked this . . . tireless migrant.

Of the twelve small pockets of forest in Mexico where monarchs from Canada and the United State overwinter, five were set aside as protected biospheres by a 1986 presidential decree. But on April 13, 1993, the *Christian Science Monitor* quoted Carlos Gottfried of Monarca A.C., a Mexican group dedicated to saving the monarch, as saying that the presidential decree is not enforced, and that illegal logging threatens the butterflies. Already, one of the protected sites has been completely deforested, half of another has been destroyed, and the buffer zones around the other three are being logged. Manuel Mondragon y Kalb, the undersecretary of Mexico's department of forests and wildlife, concedes that clandestine logging threatens all of the sites and admits that little has been done about it.

Lincoln Brower of the University of Florida at Gainesville, a longtime student of the monarch migration, told me that in 1992 70 percent of the overwintering monarchs in Mexico died, and that the monarch migration will collapse in only fifteen years if the illegal logging continues. In a recent telephone conversation, Brower told me that if we ignore this situation, we will lose these butterflies forever within our lifetimes.

The overwintering site near the little town of El Rosario has become a mecca for tourists. In 1993 about 50,000 tourists came and paid the price to visit the monarchs. This brought in close to half a million pesos, well over 150,000 U.S. dollars. This sum paid the wages of twenty winter employees, and the rest of the money was split among the 270 families of the El Rosario *ejido*, a communal land-ownership system.

Most of the monarchs from east of the Rocky Mountains spend the winter in Mexico, but monarchs from west of the Rockies gather to spend the winter in groves of Monterey pines or eucalyptus trees at several sites along the California coast from San Francisco south to Los Angeles. The most famous of these overwintering sites is in Pacific Grove, a small city just north of Monterey. I recently spoke to Ro Vaccaro of Pacific Grove, president of an organization called Friends of the Monarchs. She told me that knowledge of the Pacific Grove monarchs goes back as far as living memory, and that monarchs have probably wintered there since long before the first Spaniards arrived in California.

Little more than a decade ago, tens of thousands of these butterflies wintered in Pacific Grove. Vaccaro told me that in those days you could not walk through a monarch area unless you brushed flying butterflies away by waving your hand in front of your face like a windshield wiper. From the 1980s to the 1990s the number of overwintering monarchs declined precipitously. In the winter of 1993, only 500 of them appeared in Pacific Grove. In 1994, the overwintering population rose to between 5,000 and 6,000—a decided improvement over 1993, but still a long way from the tens of thousands that came only a few years previously.

A complex of factors is responsible for the statewide decline in overwintering monarch populations. Among them is a disease, known as green butt disease, that prevents adult monarchs from completely shedding the pupal skin. Another factor is the six-year drought that parched the Pacific coast states until recently. One of the most important factors is the maturing of the Monterey pines in which the butterflies rest. Large trees lose the lower branches on which the monarchs sit, and they are not being replaced by young trees because people put out the ground fires that heat the fallen cones and release the seeds. Another important factor has been the urban development that has destroyed or partly destroyed many of the overwintering habitats. Pacific Grove has a city ordinance that imposes a $1,000 fine on people who molest or interfere in any way with the overwintering butterflies. This law, however, has not kept people from cutting down the trees and building on sites where monarchs have overwintered as far back as anyone can remember. Vaccaro told me that not long ago, a hotel, ironically named Butterfly Trees Lodge, was built on a cut-over site that had once hosted overwintering monarchs, but no monarchs rest there now.

The citizens of Pacific Grove are, nevertheless, proud of their butterflies and have formed a monarch restoration committee to try to put an end to the destruction of overwintering sites and to restore old sites wherever possible. They have been planting young Monterey pines and nectar plants for the monarchs.

For the past fifty-four years, on the second Saturday in October, the school children of Pacific Grove have staged a parade to welcome the monarchs back. The school bands play and the kindergarten children march in butterfly costumes with orange wings pinned to their backs. One can only hope that monarchs will winter along the California coast for a very long time to come.

Silken Cocoons

The first day of fall, September 22, passed just a few days ago. The growing season is drawing to a close, and—if they have not already done so—plants and animals will soon complete preparations for the coming winter. The leaves of deciduous trees and shrubs have begun to turn, and in a few weeks the autumn colors will reach their peak of beauty all over southern Canada and the northern United States. Not long after, these colorful leaves—these no longer functional photosynthetic factories—will fall to the ground. For a season they will serve as winter cover for insects and other small animals. Then they will be recycled by scavengers, their molecules destined to serve again in the bodies of the plants and creatures of the forest.

I once knew a Massachusetts hillside that would have taken your breath away when it was in its full autumn glory in October. Let me take you there in memory. The forested, steep upper slopes of the hill are a patchwork of blazing color. The leaves of the red maples have turned a rich red, and those of the sugar maples are a medley of red, yellow, and orange. Most of the oaks are now a warm brown. Many of them will hold their leaves throughout the winter. The scattered white birches are conspicuous with their white bark and brilliant yellow leaves, and here and there the green of a white pine punctuates the warmer hues of the deciduous trees. Beneath the trees, a few witch hazels are just coming into blossom, the last of the woody plants to put forth flowers, their twisty yellow petals an epilogue to the season of reproduc-

tion. A tumbled, lichen-covered stone wall separates the forest from the abandoned pasture that is the lower and gentler slope of the hill. From the top of the wall, an irascible red squirrel chatters at an intruding dog. Some of the slope is covered with patches of brown grass, and other parts of it are carpeted with gray-green lichens, interspersed here and there with small clumps of red-capped British soldier lichens. Near the bottom of this lower slope, a small grove of aspens grows, their golden-yellow leaves fluttering and trembling in the breeze on their flexible, straplike stems. Although the aspens seem to be separate trees, they are really a single individual, a clone of shoots, each one sprouting from the roots of another. Scattered through the abandoned field are lone red cedars, straight, columnar evergreens with foliage extending from ground level to their tapered tops and already taking on its purplish winter flush. When the snows come, their blue berries—used to flavor gin—will be food for flocks of cedar waxwings and even the occasional robin that has not gone south. Parts of the lower hillside are covered with tangles of blackberry and staghorn sumac. Lacking fruit at this time of year, the thorny, burgundy-colored stems of the blackberries will soon drop the last of their leaves. The stems of the sumacs, covered with brown velvet like the new antlers of deer, bear large, candelabra-like clusters of red berries and a mantle of scarlet leaves. There are still such hillsides in Massachusetts and other parts of New England. But the one that I once knew is now covered with a housing development.

Some plants reproduced in the spring and summer: they flowered and bore the seeds that contain the embryos that will eventually germinate to become the next generation. The shadbush, so called because it blossoms in April at about the same time that the silvery shad swim up rivers along the Atlantic coast, bore its juicy, reddish berries in June, when birds eagerly ate them and spread the seeds far and wide in their droppings. But now, in the autumn, many more plants bear fruit. Mountain ashes are hung with clusters of orange fruits that will feed cedar waxwings and perhaps even Bohemian waxwings this winter. Squirrels are busy gathering and burying acorns, hickory nuts, black walnuts, and butternuts for later use. This is the season when many grasses and other plants mature their seeds, and it is now that people in the temperate zones harvest the many grass seeds that are so important in our diets: wheat, corn, rice, rye, oats, and several others. Many mammals are also reaping their harvest. Chipmunks stuff their cheeks with provender that they will store in their underground tunnels—occasionally as much as a half bushel of nuts and grain. On the rocky slopes of mountains in western North America, pikas, which are more closely related to rabbits than rodents, clip off sprigs of vegetation and lay them out in the sun to make the hay that they will store in the rocky recesses in which they live. Even some of the insects are putting in stores for the winter. Honey bees bring in the last of their harvest, nectar from late-blossoming plants such as the various kinds of asters and goldenrods.

The autumn preparations that other insects make for winter often include the spinning of a silken cocoon. Silk is secreted by many insects, even species that do not make cocoons. Among the silk spinners are silverfish, lacewings, caddisflies, beetles, moths, butterflies, bees, wasps, ants, and even fleas. Generally speaking, silk is produced only by larval forms, although a scattering of adult insects make it, too. The variety of different ways in which insects produce silk shows us that the ability to make this material that is so useful in their lives evolved independently several times during their evolutionary history. The caterpillars of moths, for example, discharge from their mouthparts silk that is secreted by a gland that originates in the head and extends back into the thorax. A small group of insects, known as the web spinners, emit silk from glands located in their front feet. Yet another way of producing silk was evolved by the larvae of lacewings and their relatives. Their silk glands are in the insectan equivalent of the kidneys, and the silk is discharged through the anus.

The uses to which insects and some of their relatives put silk are many and varied. Larval caddisflies of some species spin silken nets that strain particles of food from flowing water. Spiders spin webs, often intricate, beautiful, and geometrically precise, that

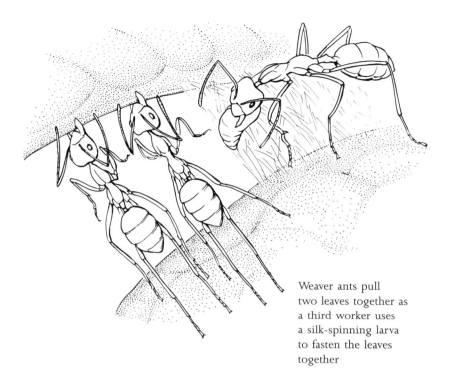

Weaver ants pull
two leaves together as
a third worker uses
a silk-spinning larva
to fasten the leaves
together

trap flying insects from the air. Caterpillars and spiders may lower
themselves to the ground on a single thread of silk that is anchored to a leaf or a twig. They may also spin a single strand of
silk that they let float off into the breeze until it is long enough
to lift them up and carry them through the air like a kite. As nicely
illustrated by Bert Hölldobler and Edward O. Wilson, adult weaver
ants use the silk secreted by their larvae to bind together the leaves
that form their arboreal nests. The workers use young larvae like
tools, holding them in their jaws and moving them back and forth
like shuttles as they tie leaves together with silk secreted by the
larvae. The naked pupae of certain butterflies are held in place on
a leaf or a plant stem by sparse silken supports that they secreted
when they were caterpillars. Hooks at the end of the pupa's
abdomen catch in a little button of silk fibers attached to the stem,
and the pupa leans back in a girdle of silk that goes around its
midsection like a window washer's safety belt. Female spiders spin

silken sacs in which they place their eggs. A few insects, bagworms and some caddisflies, for instance, spend their larval lives in silken cases that may be camouflaged with plant fragments by bagworms or with grains of sand and small pebbles by caddisflies. But people are most familiar with the silken cocoons that are spun by the caterpillars of moths. Other insects, however, also spin cocoons of silk, among them certain beetles, lacewings, ants, wasps, and fleas.

The pupae in cocoons are especially vulnerable to attack by predators in the winter, when insects and other small prey are difficult to find. At this time of year, all of the cold-blooded insect eaters, ranging from ants and praying mantises to frogs and lizards, are inactive—most of them diapausing or hibernating. Among the warm-blooded insectivores, most of the birds have gone south but some species remain behind. Woodpeckers, jays, titmice, and chickadees are with us all through the year, many of them eking out their winter diets with seeds and other vegetable foods but all foraging for what insects they can find. Some of the mammals that eat mostly plants, woodchucks and ground squirrels among them, are in hibernation, but the insect-eating species remain active throughout winter. Some of them, the mice for example, subsist largely on vegetable foods in winter, but also eat insects whenever they can get them.

The first winter that Jim Sternburg and I collected cecropia cocoons in Champaign and Urbana, we found many of them that did not contain live pupae. A few contained dry, shriveled larvae that had apparently been the victims of disease; a few contained cocoons and living pupae of the ichneumonid parasites that I described to you in an earlier chapter. But most of those that did not contain living pupae had been pierced by tiny holes, often less than one-tenth of an inch in diameter, that went through the outer and inner walls of the cocoon and into the body of the pupa itself. The viscous, semi-liquid content of the pupa had been removed through the tiny hole. Only a few scraps of tissue and the empty pupal skin remained behind.

Jim and I decided that this predator had to be a bird. All of the cold-blooded predators are inactive in winter. That left only the mammals and the birds, and none of the mammals has the

anatomical structures needed to pierce the cocoon and empty out the pupa through such a tiny hole. Many birds can peck small holes with their beaks, but only the woodpeckers can probe deeply through that hole to the farthest end of the pupa—a distance of as much as two inches. The woodpecker's tongue is barbed at its tip and can extend through a hole pecked in the trunk of a tree to reach deep into the tunnel made by an insect. It can also reach deep into a cecropia cocoon to lap out the viscous content of the pupa's body.

The circumstantial evidence was convincing, but we still had no direct evidence to prove that woodpeckers destroy cecropia pupae. We could think of two ways to get direct proof. One was to catch wild woodpeckers in the act. The other was to offer cecropia cocoons to caged woodpeckers. Eventually we managed to do both, but first we sent some cecropia cocoons to William G. George of Southern Illinois University in Carbondale and asked him to expose them to his caged hairy woodpecker. By the second day after he had wired a cocoon to the inside of the cage, the woodpecker had punched a hole in it and had hollowed out the pupa through that small hole.

It took much longer to catch a wild woodpecker in the act. That winter we watched for woodpeckers as we collected cocoons. A few times we found cocoons that had been so recently attacked that icicles of cecropia blood hung from the hole that the woodpecker had made in the wall of the cocoon. Eventually, Jim Sternburg saw a female downy woodpecker clinging to a cocoon as she punctured it. A day later we collected that cocoon and cut it open. There were the usual small holes through its inner and outer walls and through the skin of the pupa, and the pupa had been almost completely hollowed out. The next year Aubrey Scarbrough caught a hairy woodpecker in the act, and several times since then Jim and I have watched downy woodpeckers of both sexes as they methodically pecked away at cecropia cocoons.

Overwintering cocoons of cecropia and its relatives are also attacked by birds other than woodpeckers. Several birds may be involved, but we have direct evidence to implicate only two other species. During the winter of 1945, I watched a black-capped chickadee tear open a cecropia cocoon in a Connecticut wood-land. That was an unforgettable day. Early in the morning, I had a marvelous experience. As I walked around a sharp bend in the trail, I came upon a red fox posed in front of a clump of white birches as it stood in the snow like a pointer, white-tipped tail raised and one black paw lifted as it stared at me from only a few yards away.

Later in the morning, while walking in a copse of young birches, I heard a sharp scratching sound that seemed louder than it really was in the muffled silence of the snow-covered woods. I looked around to discover that the source of the sound was a black-capped chickadee hanging upside down from a cecropia cocoon on a low branch. The bird scraped at the silk until it cut a long, narrow slit through the outer and inner walls of the cocoon. Then it plunged its bill into the pupa to devour the soft tissues. The whole process took about twenty minutes.

Just a few weeks ago, I was out birding with Ellis MacLeod, friend and fellow professor in the University of Illinois Department of Entomology. We came upon a tufted titmouse that clung to a polyphemus cocoon as it pecked into its top end. The cocoon hung from a hawthorn twig by a thin thread. After a few minutes the titmouse abruptly gave up and flew off. We then collected the cocoon and discovered that it contained only the dry and shriveled remains of a larva. Apparently, the titmouse had kept pecking until it discovered that the cocoon contained nothing edible. But we have never seen a cocoon that had not contained an intact pupa when it was attacked by a woodpecker. It looks as if woodpeckers, unlike titmice, can tell if a cocoon contains a living pupa before they make a hole in it—perhaps by perceiving the solid weight of the living pupa as it slides about when they first shake the cocoon by landing on it or pecking at it.

An as yet unidentified kind of bird, possibly the blue jay or the white-breasted nuthatch, stuffs morsels of food through the exit valve of cecropia cocoons into the space between the outer and inner walls. From 1966 to 1968, Jim Sternburg, Aubrey Scarbrough, and I collected about 2,000 cecropia cocoons in Champaign and Urbana; at least 54 of these cocoons, about 3

percent, had foreign objects jammed into their valves. In one cocoon, the valve was so completely blocked that the moth died because it could not push its way out. These 54 cocoons contained seventy-five different objects; twenty-six unshelled sunflower seeds, twenty-one kernels of corn, eleven unidentified pieces of nutmeat, eight shelled sunflower seeds, two half peanut kernels, two whole acorns, two pieces of shelled acorn, one entire peanut kernel, one piece of Brazil nut, and one dead earthworm!

White-footed and deer mice also eat cecropia pupae, but they penetrate the walls of the cocoon in a different way than do birds. These mice chew a large, oval hole in the cocoon, a hole that may be as much as two inches long and an inch wide. Through this hole they remove the pupa and take it away. Sometimes they carry away the whole intact cocoon. One of the graduate students in our department, Lloyd Davis, once found the nest of a white-footed mouse that had a pile of four cecropia cocoons stored next to it. Nest and cocoons were tucked away in a dry nook under a log lying on the ground. Three of the cocoons had been opened, but the other one was still intact. We have never caught a mouse in the act of opening a cocoon under natural conditions, but, as you will see in the last chapter, we have found that both white-footed and deer mice will readily open cecropia cocoons in the laboratory and will greedily eat the pupae.

By this time in September the last of the cecropia caterpillars have just finished spinning their cocoons and are about to molt to the pupal stage. The behaviors that led them to suitable locations for spinning a cocoon are stereotypical: all individuals of the species perform essentially the same behaviors following genetically determined programs. Unless frustrated by atypical environmental circumstances, these stereotypical behaviors of the caterpillars lead them to locate their cocoons in relatively safe places where they are hidden from view.

When a cecropia caterpillar has grown to full size and is ready to seek a place in which to spin its cocoon, it stops eating, crawls to the underside of a branch, and clings there with its abdomen drooping down as it evacuates its intestines. It then moves to the upper side of the branch, and crawls down toward the ground, waving its head from side to side in a "swing-swing" movement as it lays down a zigzag trail consisting of a single strand of silk that clings to the branch. Caterpillars that feed on shrubs usually move down to spin their cocoons near ground level among the stems of the very shrub on which they fed or on some other nearby shrub. Caterpillars that feed in trees usually descend to spin their cocoons elsewhere. Generally they move to a nearby shrub, sometimes an evergreen, to construct their cocoons near ground level. Cocoons that are spun in such ground-level locations are soon covered by the falling leaves of autumn and are very well hidden.

Cecropia caterpillars that have made their way to the ground from the branches of a tree—say, a silver maple growing on a suburban lawn—can be seen to pause, lift their heads, and peer from side to side as they crawl along across the ground, as if looking for a likely spinning site. If a shrub is nearby, the caterpillar will head straight toward it. If no shrub is in sight, the myopic caterpillar continues to crawl away from the tree, often stopping to lift its head as if looking for something.

A simple experiment that Aubrey Scarbrough did suggests that this is exactly what the caterpillar is doing. On a wide suburban lawn bare of shrubs, he watched prespinning caterpillars as they crawled across the lawn after descending from a tall silver maple. As each caterpillar crawled along, he placed a black cardboard silhouette shaped like a shrub about three feet to one side of its anticipated path, so that its shadow was cast *away* from the caterpillar. When the insect came abreast of this two-dimensional model, it stopped, lifted its head, and turned toward it.

Thirty-six of forty caterpillars so tested changed direction and made their way to the silhouette. Their behavior leaves little doubt that they were trying to get a visual fix on a potential cocoon-spinning site. Caterpillars that crawled thirty feet or more from the base of the tree without finding a spinning site usually turned around and climbed back up into the tree to spin a cocoon in its lower branches, but no one knows if they followed the trail of silk that they had laid down on their outward trip.

Once the caterpillar decides on a precise location, it apparently uses visual cues to judge the suitability of that site. In a dark and shaded nook, the cocoon will probably be hidden from the view of predators. In a bright and sunny site, it will almost certainly be exposed to view. Thus it came as no surprise to our little research group that cecropia caterpillars search for spinning sites only during daylight. Most of them begin their search from four to eight hours after sunrise, and almost all of them have chosen a site and have begun to spin their cocoons by about two hours before dark. The few caterpillars that do not locate a spinning site before dark climb up onto a plant stem and cling there as they wait to resume their search with the return of daylight the next day.

The intricate cocoon spun by a cecropia caterpillar is the tangible and decipherable record of a complex behavior. There are three layers of silk in a completed cocoon: a tough, leathery outer envelope, a thicker and even tougher inner envelope within which the pupa rests, and a fluffy intermediate layer that suspends the inner envelope within the outer envelope. When the caterpillar arrives at its spinning site, it continues the swing-swing movements of its head, but now—because it is no longer crawling—this results in the laying down of a sheet of silk rather than a trail. This sheet, attached to a stem or a twig, is the foundation of the cocoon, which firmly attaches the outer envelope to the woody twig or stem of the plant on which the cocoon is spun. If leaves are nearby, one or more of them is usually incorporated in the outer envelope. Even after an incorporated leaf has dried and crumbled away, the impression of its veins may still be visible on the surface of the cocoon. After the outer envelope has been completed, the caterpillar spins the intermediate fluffy layer that supports the inner envelope, the last part of the cocoon to be spun. In the front end of each of these envelopes, it constructs a valve that will permit the newly emerged moth to push its way out of the cocoon the following spring.

Some insects spin cocoons that appear to have little or no function. Their walls are so thin and flimsy that the pupa is readily visible. Other insects, cecropia and its close relatives promethea and polyphemus, for instance, spin cocoons whose thick, leathery walls protect the pupa against the hordes of predators that would be delighted to eat it if they could only get at it. The great majority of potential predators lack the anatomical equipment or the behavioral capability to breach the walls of these sturdy cocoons. A few, however, such as the woodpeckers and mice that I have already mentioned, are able to penetrate them. But natural selection is a continuing process. The arms race of attacks countered by new defenses goes on, and cecropia, polyphemus, promethea, and many other insects have responded to these few predators by evolving ways to avoid their attacks at least sometimes.

The caterpillars of cecropia, promethea, and polyphemus have each evolved their own way of decreasing the probability that they will be destroyed by a bird or a mouse during the winter. As you have already seen, cecropia caterpillars usually descend to ground level to spin their cocoons among the stems of a shrub, where they are eventually covered by fallen leaves. Such well-hidden cocoons almost always escape the notice of woodpeckers, but, as we will see later, they are sometimes attacked by mice.

Promethea caterpillars use a completely different strategy to foil the attacks of both woodpeckers and mice on their cocoons. They do not leave the tree on which they fed, usually a sassafras or a wild black cherry. Instead, they move to the tip of a twig, often high in the tree, where they securely fasten a leaf to the tree by means of a strong silken strap that surrounds the leaf stem and attaches it firmly to the wooden twig. The caterpillar then spins its cocoon within the folded blade of the leaf. Even if the leaf crumbles away during the winter, the cocoon will not fall because it remains firmly attached to the twig. Although white-footed mice do climb onto the lower branches of trees, you very seldom find promethea cocoons that have been attacked by them. It may be that the mice are reluctant to venture out onto the thin, flexible, terminal twigs from which the cocoons hang. Furthermore, you rarely find promethea cocoons that have been penetrated by woodpeckers or other birds. We think that the woodpeckers seldom attempt to perch on the cocoon itself because it is too small for them to gain a secure foothold. Even if a woodpecker finds some other perch from which it can reach the cocoon, it will not be able to penetrate the cocoon unless it can manage to peck directly down into the valve. The force of such a blow is directed in line with the supporting strap and will not cause the cocoon to swing away, but a peck directed at any other part of the cocoon will cause it to swing away on its flexible strap and will thus not result in penetration.

Although promethea cocoons are very seldom attacked by woodpeckers or mice, Jim Sternburg and I did find an exceptional circumstance in which a large number of them, all within a few yards of each other, had been penetrated by woodpeckers. Along the side of a country road in Indiana, a snowplow had thrown up a bank of snow that buried the lower branches of several wild black cherries. As the snow melted away, some promethea cocoons that hung from the twigs of these trees were partly exposed but still held in place by the frozen snow. Woodpeckers, always seeking and always opportunistic, were able to perch on the ice and penetrate the cocoons because the cocoons could not swing away.

Polyphemus cocoons have evolved yet another strategy to minimize the exposure of their cocoons to the attacks of mice and birds. This insect spins its cocoon near the tip of a twig of the tree on which it fed, usually an oak, a hawthorn, a birch, or a maple. Like promethea cocoons, polyphemus cocoons are generally wrapped in a leaf, but they differ from promethea cocoons in that the leaf is only loosely attached to the twig, usually by just a few strands of silk. The great majority of these cocoons fall to the ground in autumn with their attached leaves. The leaf may even act as a sail by which the wind can blow the cocoon away from beneath the tree from which it fell. At any rate, the cocoons tend to become randomly scattered among the fallen leaves on the ground. Here they are hidden from woodpeckers and are probably seldom found by mice because they are so well dispersed among thousands of dead leaves. There is no doubt that woodpeckers would attack polyphemus cocoons if they could find them; the majority of the few cocoons that do not fall from the tree are indeed destroyed by woodpeckers.

People have had a long, delightful, and profitable association with domestic silkworms and the silk of their cocoons. The silkworm of commerce is actually an insect, the caterpillar of a moth that is not too distantly related to cecropia and the other giant silkworms. This insect has been domesticated for millennia and it no longer exists in its natural, wild state. During its domestication, people selected it for the quantity and quality of the silk that it spins and for behavioral characteristics that make it easy to handle in captivity. Consequently, no one knows how it behaved before it was domesticated. Silkworm moths have wings but do not fly, and the caterpillars seldom move as long as food is available.

It is doubtful that such a lethargic insect could survive in nature. The caterpillars would be in the same place generation after generation, and as their population grew they would eventually deplete their food supply. When they are fully grown, the caterpillars spin their cocoons in the closest available site and, depending upon the variety of silkworm, the silk of the cocoon is either bright yellow or white. In nature, the pupae in cocoons that are not camouflaged and that are so conspicuously placed would be easy prey for birds, mice, and other predators. Female silkworm moths have also lost the instinct to lay their eggs on the plants that their larvae will eat. They will lay them almost anywhere, usually on paper purposely laid down for them in commercial cultures. Clearly, the silkworm has become completely dependent upon humans for its very existence.

People still culture silkworms much as they did a thousand years ago, although there have been some modern innovations. The caterpillars are usually fed the leaves of white mulberry and, as you read earlier, they will prosper and produce silk in good quantity and quality only if they are fed the leaves of this tree or those of a few other trees in the same family. During the winter silkworm eggs are in diapause and are stored in a cold place. In the spring they must be kept at a temperature which ensures that hatching will coincide with the unfurling of the leaves on mulberry trees. Large groups of the caterpillars are kept on open trays, but they rarely escape because they seldom crawl for more than a short distance. Throughout their lives they are supplied with chopped mulberry leaves; the larger the larvae grow, the more coarsely the leaves are chopped. The trays are frequently cleaned of old leaves and excrement.

Like most caterpillars, silkworms are voracious feeders. They have to be. From the time it hatches out of the egg at an average weight of only 1 milligram until it attains its maximum weight, an average of about 3,580 milligrams (about thirteen-hundredths of an ounce) as a full-grown caterpillar, a silkworm increases its weight by an astonishing factor of 3,580 times. During the twenty-eight days that it takes a caterpillar to mature, it will have eaten well over 15,000 times its own weight as a hatchling. But during the first twenty-two days of their lives, these caterpillars feed so modestly that an amateur silkworm raiser can be lulled into a false confidence in his ability to supply mulberry leaves to his pets. But during the last six days of its life a caterpillar will eat five times as much as it did during the first twenty-two days. In other words, it consumes well over 80 percent of its lifetime intake of food during the last 20 percent of its life as a caterpillar. When the full-grown caterpillars have finished eating, bunches of twigs or loose bundles of straw are placed on the trays. Here the caterpillars spin the cocoons that are then processed as described below. A group of caterpillars must eat about sixty pounds of mulberry leaves to produce a single pound of silk.

The origin of the domesticated silkworm is lost in the distant past, but it seems clear that the use of silk originated in China during the Stone Age. According to the Cambridge Encyclopaedia of China, traces of cocoons and silk fabrics were recovered at a late Stone Age (Neolithic) living site in China that dates to almost 7,000 years ago. But this early origin of silk has been lost to Chinese tradition, which holds that the usefulness of silk was discovered by the Empress Hsi Ling-Shi in 2640 B.C., over 2,300 years after the Neolithic date. She is also credited with discovering the process of unwinding the single strand of silk that forms the cocoon. According to legend, when Hsi Ling-Shi retrieved a cocoon that had accidentally fallen into hot water (sometimes said to have been a cup of tea) it came out dangling from a single fiber of silk that could be unwound from the cocoon.

To this day, essentially the same process is used to unwind cocoons. First, the cocoons are cleaned by picking off bits of leaf and other debris that clings to them. Then they are steamed for a few minutes to kill the pupae within them. (The silk would be ruined if a moth emerged from the cocoon). Finally, they are immersed in boiling water to loosen the fiber by dissolving the outer layer of gelatinous sericin that coats the inner core of fibroin and binds all of the silk in the cocoon together. When the sericin has been removed, the end of the fibroin thread can be found and attached to a machine-driven reel that will unwind all 2,000 to 3,000 feet of the fiber. At the end, when all of the silk has been reeled up, nothing is left behind but the naked pupa. Two or more single fibers may be twisted ("thrown") together to form threads of varying thickness that are ultimately woven into the silken fabrics that are treasured all over the world. The pupae are not wasted. The Chinese have recipes for preparing them as food for humans.

Silk is a luxurious fabric, even a sensuous fabric. Its feel on the skin is a caress. It can be nearly transparent, and it clings and reveals the curves of the human body. As early as the middle of the first century A.D., Roman writers, all men, complained that wearing silk "rendered women naked in the street." In 77 A.D., the Roman encyclopedist Pliny the Elder wrote of silkworms and silk in his *Naturalis Historia*: "They weave webs like spiders, producing a luxurious material for women's dresses, called silk. The process for unraveling these and weaving the thread again was first invented in Cos by a woman named Pamphile, daughter of Plateas, who has the undeniable distinction of having devised a plan to reduce women's clothing to nakedness." Pliny here seems to refer to the unraveling and reweaving of silk fabrics obtained from the Orient. This process produced a much thinner and presumably almost transparent fabric.

He then writes of another sort of silk that he believed to be the down of leaves that is scraped together by naked butterflies and wound into a fleece that they wrap around their bodies to keep themselves warm in winter. He goes on to report that tufts of this "wool" are plucked off and "softened with moisture and then thinned out into threads with a rush spindle. Nor have even men been ashamed to make use of these dresses, because of their lightness in summer: so far have our habits departed from wearing a leather cuirass that even a robe is considered a burden! All the same we so far leave the Assyrian silk-moth to women." He is, of course, mistaken in his biological notions about silk, but his sociological observations have the ring of truth.

The Chinese successfully guarded the secret of silkworm culture for thousands of years. The penalty for smuggling either silkworm eggs or mulberry seeds out of the country was death. While maintaining their monopoly, they traded silk fabrics to the ancient Greeks and the Romans. Pliny, Ovid, and Virgil all wrote about silk. Silk moved westward from China to the Mediterranean along the legendary Silk Road, while commodities such as amber, silver, gold, and wool moved eastward to China. According to John Feltwell in *The Story of Silk,* the Silk Road was established at least 4,000 years ago and ran for over 3,000 miles from the Pacific coast of China to Tyre on the east coast of the Mediterranean. From there boats carried silk to Greece, Rome, and other parts of ancient Europe.

The Chinese maintained their monopoly on silk culture for thousands of years, but lost it in 552 A.D. when two Christian monks from Persia, encouraged by the Byzantine emperor Justinian, smuggled silkworm eggs out of China in hollow staffs and carried them to Constantinople (modern Istanbul), then the capital of the Byzantine Empire. Silk culture thrived in Byzantium and eventually spread to such European countries as France, Italy, and England. Silk production was even introduced into the United States, but ultimately failed there. Silk has had its ups and downs with changes in fashion and especially with the introduction of rayon ("artificial silk") in 1910. But silk is still—and will probably always be—treasured for its luxurious and inimitable qualities, and to this day the silk industry persists in over thirty countries. Today the major producers of silk are China (still the leading producer), Japan, India, South Korea, Brazil, and Uzbekistan, formerly in the Soviet Union. As John Feltwell says, the future of silk production looks bright.

Winter

Now that winter is here, we have come full circle in the seasons of the biological year. Spring burgeoned into summer, summer ripened into autumn, and autumn has now faded into winter. The look of the land has changed again—much more drastically than it did in late summer and fall. In much of our area, the ground is covered with snow, and lakes and ponds are frozen. The predominant colors are white, gray, and brown. Gone are the greens of summer and spring and the gay yellows and oranges of autumn. Few herbaceous plants are to be seen; most of them are represented only by seeds or perennial root systems in the soil. Almost all of the broad-leaved trees have dropped their leaves to the ground, where they serve as cover for many kinds of dia-pausing insects; hemlocks, pines, and firs remain green, but even a few of the conifers have lost their needles, the tamaracks in the north and the bald cypresses in the south.

Winter is the frugal season. Food is scarce. Squirrels are drawing on the stores of nuts that they buried in the fall. In their ice-covered ponds, beavers retrieve branches of aspen, willow, and birch from which they gnaw the bark—branches that they cut and anchored in the mud earlier in the year. Here and there, a belted kingfisher that did not go south dives to catch a minnow from an unfrozen area along a stream or from the warm water below the outlet of a power plant. Human anglers chop holes in the ice and use maggots from goldenrod galls as bait to fish for bluegills. The chorus of bird songs that we heard from forest and field in spring and summer is now silent. But crows and jays still call, and from a nearby woodland we hear the nasal "yanks" of a white-breasted nuthatch and the loud, slow, and steady chop, chop, chop of a crow-sized pileated woodpecker as it digs in the wood of a dead tree for beetle grubs, each stroke of its chisel-like bill knocking out a three- or four-inch chip. At the edge of the woods, a smaller downy woodpecker, about the size of a sparrow, perches on a goldenrod stalk as it pecks at a gall to get the same kind of maggot that the ice fishers use for bait. Other birds also scrounge for food. Brown creepers describe a helical path up the trunk of a tree as they pick tiny insect eggs from crevices in the bark. Flocks of black-capped chickadees, often accompanied by titmice, nuthatches, and downy woodpeckers, move through the woods as they forage for food. In open country, mixed flocks of horned larks, lapland longspurs, and an occasional snow bunting search for seeds in bare areas along the sides of roads cleared by snowplows. In a shrubby area near the woodland edge, a sleek and altogether charming white-footed mouse moves about beneath the snow as it searches the ground litter for overwintering insects, seeds, and other morsels of food.

We have also come full circle in the life history of cecropia. In the first chapter we became acquainted with a cecropia pupa that had survived the winter in a cocoon hidden away in a dense bush beneath a birch tree. It was then the end of March—after the pupa had experienced the long winter chill that primed it to come awake in response to the first warm days of spring. As you read, the infinitesimal spark of life in the diapausing pupa was fanned into a tiny flame as metamorphosis began. About two months later, the adult moth emerged from her cocoon, attracted a mate, and then went on to lay her eggs. Now that winter is here, her surviving descendants and those of all of the other cecropias of her generation are nestled in their cocoons. This is the best opportunity of the year to find these insects. Although most of them are hidden near ground level, they can be found with diligent searching.

During three winters in the late 1960s, Jim Sternburg, Aubrey

A pileated woodpecker has just exposed a large beetle grub under the bark of a tree

Scarbrough, and I collected cecropia cocoons in the twin cities of Champaign and Urbana. We were far more successful than we thought we would be. We found over 2,000 cocoons during those three years. Although some of our colleagues thought that we were being overly meticulous, we kept track of the street address at which we found each cocoon. It's a good thing that we did. That information gave us an insight into cecropia life that we would otherwise have missed. When we stuck pins representing the cocoons into a map, we saw that their abundance varied between neighborhoods of different ages. We found very few cocoons in old neighborhoods with mature shade trees that had been settled in the late nineteenth and early twentieth centuries. In aerial photographs of these old areas, the crowns of the trees are seen to overlap and largely obscure the roofs of buildings. By contrast, cocoons were abundant in newly settled suburban areas. Since these areas are all built on former crop fields, the shade trees are small and do not at all obscure the outlines of buildings in aerial photographs. In areas of intermediate age, with the outlines of buildings only slightly obscured in aerial photographs, cecropia cocoons were moderately abundant. The number of cocoons found during the three years differed tremendously between these three types of neighborhoods. In old areas we found only 136 cocoons, less than 6 percent of the total. But in newly built suburban areas we found a whopping 1,771, over 76 percent of the total. In intermediate areas we found 420 cocoons, about 18 percent of the total.

As I wrote earlier, we think that cecropia is a "fugitive species" whose population is constantly shifting to disturbed areas such as abandoned crop fields that are in an early stage of the succession that will return them to mature forest, the stage in which sapling trees and shrubs have appeared. Later in the succession, when the area has become a mature forest, cecropia populations will be much smaller.

It occurred to us that this insect's scarcity in old residential areas and its abundance in newly settled suburban areas might reflect its preference for early stages in the natural succession. After all, an old residential area with its mature shade trees is the "climax stage" of succession in the urban forest. And newly settled areas at the fringe of the urban area, with their shrubs and sapling shade trees, are much like an early stage in a natural succession. Indeed, during the fifteen years of our studies of cecropia in the Champaign-Urbana area, Jim Sternburg and I have seen a major shift in the urban cecropia population. Suburban areas that had large populations in 1965 had only small populations in 1980. By that time the trees in these areas were nearing maturity, and these areas

were much further along in the urban succession. But the suburban areas that were newly developed in 1980, on land that had been crop fields in 1965, had large cecropia populations.

Although cecropia is much less abundant in rural than in urban areas (more about this later), it does occur in a few rural areas that mimic an early stage in the natural succession, but is almost absent from mature woodlands. In one year we found only 1 cecropia cocoon after a thorough search of two mature woodlands. But in roadside growths of saplings, shrubby willows, and red osier dogwood shrubs we found 92 cocoons. To keep things in perspective, we found 980 cocoons in Champaign and Urbana during the same year, most of them in newly built suburban areas.

Cecropia has become an insect that associates with humans. To many of us, this spectacular insect is a welcome guest. It is never abundant enough to seriously damage young shade trees or shrubs, and it is a delight to those of us who cherish nature in all of its shapes. Our observations raise two questions, the answers to which are important to understanding cecropia's distribution in cities and towns. First, why is this species so much more abundant in cities and towns than it is elsewhere? Second, why is it common in recently built suburban areas but almost absent in old residential areas? Cecropia's abundance in cities and towns is nothing new. As long ago as 1899, this moth was reported to be especially abundant in the cities of New Jersey and Long Island. In her *Moths of the Limberlost*, Gene Stratton Porter wrote that she found many cecropia cocoons in Indianapolis but far fewer in rural areas of Indiana. We do not have all of the answers to these questions, but we do have some answers that account at least partially for the peculiarities of cecropia's distribution.

The first question can be more specifically put by asking why cecropia is so much more abundant in suburban areas than it is in rural areas that mimic a comparably early stage in the ecological succession. One answer to this question is simply that the species of mice that prey upon cecropia cocoons in rural areas are replaced in suburban areas by that long-time associate of humans, the common and pestiferous house mouse. Although we generally think of house mice as pests in our homes, they also occur outdoors in urban and suburban areas. When we trapped in the front yards of suburban homes, we caught 102 mice—95 of them house mice and only 7 of them the native white-footed or deer mice, the species that so commonly prey on the pupae in cecropia cocoons in rural areas. The abundance of these mice was reversed when we trapped in rural areas. There we caught only 3 house mice but 39 white-footed mice and 20 deer mice.

Behavioral studies done in our laboratory showed that house mice never attack cecropia cocoons, but that the two native species of mice regularly do so. We kept 12 of the deer mice, 25 of the white-footed, and 25 of the house mice in laboratory cages until they had settled down and regularly ate Purina Rat Chow and had built nests in which to sleep. Then we presented each of these mice with an unopened cecropia cocoon. All but 2 of the deer mice and all of the white-footed mice eventually chewed open the cocoon, removed the pupa, and thereafter preferred to eat cecropia pupae rather than rat chow. But none of the house mice opened a cocoon. They readily ate naked cecropia pupae that we made available to them, they ripped silk off the cocoons to incorporate it in their nests, but they would not chew open a cocoon to obtain a pupa even when faced with starvation. Opening cocoons is clearly not in the behavioral repertoire of house mice.

In both suburban and rural areas, some cecropia cocoons are spun above ground level on the branches of shrubs and trees. Most of these cocoons are in plain sight after the leaves have fallen in autumn, and during the winter the pupae in about 90 percent of them fall prey to woodpeckers in both rural and suburban areas. But the fate of cocoons spun close to ground level is very different in suburban and rural areas. Almost all of the pupae in suburban cocoons spun low among the stems of shrubs survive the winter. Out of the hundreds of cocoons that we found near ground level in urban areas, less than a dozen had been opened by mice. In rural areas, however, about 62 percent of the cocoons that had been spun near ground level had been attacked by mice. Most of them had been opened where they had been spun and the pupae had been removed. There is no doubt that the cocoons had been

A white-footed mouse chews a hole in a polyphemus cocoon to get at the pupa

opened by white-footed or deer mice, since the damage seen on them was identical to damage done by these mice in the laboratory. But in a few instances, almost the whole cocoon had been removed and nothing remained behind except shreds of the silk that had moored it to a stem. These cocoons were probably stored near the mouse's nest until the pupae were needed for food.

The remaining question is why cecropia is so much scarcer in old residential areas than it is in newly built suburban areas. Again, we do not have all of the answers, but we can point to one of the major factors. While cecropia caterpillars are fairly safe from insect-eating birds in new suburban areas, in old residential areas, as well as in mature forests, they are eaten by several species of birds that are less common in newly developed suburbs. In general, birds are relatively scarce in new suburbs, but in old residential areas they are fairly abundant. Among the birds of old areas are several species that feed almost exclusively on insects or add them to their partly vegetarian diets. They include common grackles, blue jays, house wrens, catbirds, cardinals, brown thrashers, and mockingbirds. At one time or another, we have seen all but the grackle and the wren eating cecropia caterpillars.

Let me return to the cecropia moth that we met in the first chapter, the one that had spun her cocoon in the dense bush beneath the birch tree. It would be instructive to know her fate and the fate of her descendants, but no one has ever kept track of a cecropia moth and her offspring through the course of a year.

That is almost certainly an impossible task under natural conditions. Nevertheless, from what is known, I can construct a hypothetical scenario that approximates the truth.

By virtue of good genes and good luck, our cecropia moth left behind more than enough progeny to replace herself and her mate. She laid well over 200 eggs before she was snatched out of the air by a bat on the fourth night of her life. Some of her eggs never hatched; they were eaten by foliage-gleaning birds such as migrant warbling vireos. Only a few dozen of the caterpillars that hatched survived the gauntlet of predators and disease to grow to full size and construct cocoons. Immediately after they spun their cocoons, several of them were devoured by parasitoids that had been lurking in their bodies and would now use their cocoons as winter shelters. A few others were only then parasitized by wasps that, attracted by the odor of fresh silk, pierced the cocoon to place an egg in the larva's body. During the coming winter, the pupae in several other cocoons will be killed by woodpeckers. Only four of the surviving pupae will make it through the winter to emerge as moths in the following spring. And one of these moths, before it can mate, will be eaten by a foraging brown thrasher. The surviving three will all mate. The two females among them will lay many eggs, and the lone male will inseminate two females. The moth from beneath the birch did well indeed. Not only did she replace herself and her mate, but she left behind one extra reproducing descendant. Her genes will be well represented in the generation of her grandchildren.

Just a few lines above I wrote that this female was successful by virtue of good genes and good luck. Her good genes ensured that her programmed behaviors and those of her offspring worked optimally so as to increase the chances for survival. She placed her eggs on favorable plants, her larval offspring spun cocoons that were well hidden, and scores of other behaviors were performed as dictated by millennia of natural selection. Her good luck was that her mother had laid the egg from which she hatched on a leaf of a white birch that grew in front of a house in a recently built suburban area. Directly beneath the birch grew a juniper shrub in which she spun her cocoon. If our female cecropia had

found herself in a rural area, she might not have reproduced so successfully.

Cecropia populations in rural areas or old urban neighborhoods generally remain low from year to year, because parents usually manage to do no more than replace themselves from generation to generation. But in suburban areas populations tend to grow, often at a rapid rate, because parents often more than replace themselves. Most of the over 200 progeny left by our suburban female perished before they reached the reproductive state, but 3 of them did survive to reproduce. We tend to focus on the hundreds of progeny that fell by the wayside, but, from the point of view of population growth, we should focus on the 3 that survived. At this rate of replacement, the population will increase by 50 percent each year. In actuality, Jim Sternburg and I found that a suburban cecropia population in Champaign and Urbana increased by an average of over 80 percent per year over a three year period. Where we found 285 cocoons the first year, we found 506 the second year and 958 the third year. Such population growth cannot, of course, continue indefinitely. It will eventually stop. But that is another story.

Selected Readings

First Things

Attenborough, D. 1979. *Life on Earth*. Boston: Little, Brown.

Borror, D. J., and R. E. White. 1970. *A Field Guide to the Insects*. Boston: Houghton Mifflin.

Colinvaux, P. A. 1978. *Why Big Fierce Animals Are So Rare: An Ecologist's Perspective*. Princeton: Princeton University Press.

Covell, D. V., Jr. 1984. *A Field Guide to the Moths of Eastern North America*. Boston: Houghton Mifflin.

Darwin, C. 1964. *On the Origin of Species: A Facsimile of the First Edition*. Cambridge, Mass.: Harvard University Press.

Knutson, R. M. 1974. Heat production and temperature regulation in eastern skunk cabbage. *Science*, 186:746–747.

Linsenmaier, W. 1972. *Insects of the World*. Translated from the German by L. E. Chadwick. New York: McGraw-Hill.

Romoser, W. S., and J. G. Stoffolano, Jr. 1994. *The Science of Entomology*. Dubuque, Iowa: Wm. C. Brown.

Ross, H. H., C. A. Ross, and J. R. Ross. 1982. *A Textbook of Entomology*. New York: John Wiley and Sons.

Sternburg, J. G., and G. P. Waldbauer. 1969. Bimodal emergence of adult cecropia moths under natural conditions. *Annals of the Entomological Society of America*, 62:1422–1429.

Villiard, P. 1969. *Moths and How to Rear Them*. New York: Funk and Wagnalls.

Waldbauer, G. P., and J. G. Sternburg. 1973. Polymorphic termination of diapause by cecropia: genetic and geographical aspects. *Biological Bulletin*, 145:627–641.

————1979. Inbreeding depression and a behavioral mechanism for its avoidance in Hyalophora cecropia. *The American Midland Naturalist,* 102:204–208.

The Most Successful Animals on Earth

Cumber, R. A. 1953. Some aspects of the biology and ecology of bumble-bees bearing upon the yields of red-clover seed in New Zealand. *New Zealand Journal of Science and Technology,* ser. B, 34:227–240.

DeBach, P. 1964. *Biological Control of Insect Pests and Weeds.* New York: Reinhold Publishing.

Doutt, R. L. 1958. Vice, virtue, and the vedalia. *Bulletin of the Entomological Society of America,* 4:119–123.

Erwin, T. L. 1983. Tropical forest canopies: the last biotic frontier. *Bulletin of the Entomological Society of America,* 29:14–19.

Hanski, I., and Y. Cambefort, eds. 1991. *Dung Beetle Ecology.* Princeton: Princeton University Press.

Hocking, B. 1968. *Six-Legged Science.* Cambridge, Mass.: Schenkman Publishing.

May, R. M. 1988. How many species are there on earth? *Science,* 241:1441–1449.

McGregor, S. E. 1976. *Insect Pollination of Cultivated Crop Plants.* Agricultural Handbook no. 496. Washington, D.C.: U.S. Department of Agriculture.

Montgomery, B. E. 1951. The status of bumble bees in relation to the pollination of red clover in New Zealand. *Proceedings of the Sixth Annual Meeting of the North Central States Branch of the American Association of Economic Entomologists,* 51–55.

Poinar, G. O., Jr. 1992. *Life in Amber.* Stanford: Stanford University Press.

Sabrosky, C. W. 1952. How many insects are there? In *Insects: The Yearbook of Agriculture.* Washington, D.C.: U.S. Department of Agriculture.

Waterhouse, D. F. 1974. The biological control of dung. *Scientific American,* 230(4):100–109.

Finding and Courting a Mate

Alexander, R. D., and T. E. Moore. 1962. The evolutionary relationships of 17-year and 13-year cicadas, and three new species (Homoptera,

Cicadidae, *Magicicada*). *Miscellaneous Publications of the Museum of Zoology, University of Michigan*, 121:1–59.

Bristowe, W. S. 1976. *The World of Spiders*. London: Collins.

Brower, L. P., J. vz. Brower, and F. P. Cranston. 1965. Courtship behavior of the queen butterfly, *Danaus gilippus berenice* (Cramer). *Zoologica*, 50:1–39 + 7 plates.

Butenandt, A., R. Beckmann, D. Stamm, and E. Hecker. 1959. Über den Sexual-Lockstoff des Seidenspinners *Bombyx mori*: Reindarstellung und Konstitution (On the sexual attractant of the silkworm *Bombyx mori*: purification and structure). *Zeitschrift für Naturforschung*, 14:283–284.

Chapman, J. A. 1954. Studies on summit-frequenting insects in western Montana. *Ecology*, 35:41–49.

Comstock, J.H. 1950. *An Introduction to Entomology*. 9th ed. Ithaca: Comstock Publishing.

Cowan, F. 1865. *Curious facts in the History of Insects*. Philadelphia: J. B. Lippincott.

Cutler, W. B., G. Preti, A. Krieger, G. R. Huggins, C. R. Garcia, and H. J. Lawley. 1986. Human axillary secretions influence women's menstrual cycles: the role of donor extract from men. *Hormones and Behavior*, 20: 63–473.

Dethier, V. G. 1992. *Crickets and Katydids, Concerts and Solos*. Cambridge, Mass.: Harvard University Press.

Graham, C. A., and W. C. McGrew. 1980. Menstrual synchrony in female undergraduates living on a co-educational campus. *Psychoneuroendocrinology*, 5:245–252.

Kirby, W., and W. Spence. 1846. *An Introduction to Entomology*. 6th London ed. Philadelphia: Lea and Blanchard.

Lloyd, J. E. 1975. Aggressive mimicry in *Photuris* fireflies: signal repertoires by femmes fatales. *Science*, 187:452–453.

Magnus, D. B. E. 1958. Experimental analysis of some "overoptimal" sign-stimuli in the mating behaviour of the fritillary butterfly *Argynnis paphia* L. (Lepidoptera: Nymphalidae). *Proceedings of the Tenth International Congress of Entomology*, 1956, 2:405–418.

Maier, C. T., and G. P. Waldbauer. 1979. Diurnal activity patterns of flower flies (Diptera: Syrphidae) in an Illinois sand area. *Annals of the Entomological Society of America*, 72:237–245.

——— 1979. Dual mate-seeking strategies in male syrphid flies (Diptera: Syrphidae). *Annals of the Entomological Society of America*, 72:54–61.

Roth, L. M. 1948. A study of mosquito behavior: an experimental laboratory study of the sexual behavior of *Aedes aegypti* (Linnaeus). *American Midland Naturalist,* 40: 65–352.

Thornhill, R., and J. Alcock. 1983. *The Evolution of Insect Mating Systems.* Cambridge, Mass.: Harvard University Press.

Tinbergen, N. 1972. *The Animal in Its World: Explorations of an Ethologist.* Vol. 1, *Field Studies.* Cambridge, Mass.: Harvard University Press.

Weller, L., and A. Weller. 1993. Human menstrual synchrony: a critical assessment. *Neuroscience and Behavioral Reviews,* 17:427–439.

Williams, C. M. 1946. Physiology of insect diapause: the role of the brain in the production and termination of pupal dormancy in the giant silkworm, *Platysamia cecropia. Biological Bulletin,* 90:234–243.

———— 1958. Hormonal regulation of insect metamorphosis. In W. D. McElroy and G. Glass, eds., *A Symposium on the Chemical Basis of Development.* Baltimore: Johns Hopkins University Press.

Wilson, E. O., and W. H. Bossert. 1963. Chemical communication among animals. *Recent Progress in Hormone Research,* 19:673–716.

After the Courtship's Over

Alexander, R. D., and D. Otte. 1967. The evolution of genitalia and mating behavior in crickets (Gyrllidae) and other Orthoptera. *Miscellaneous Publications of the Museum of Zoology, University of Michigan,* 133:1–62.

Beebe, W. 1947. Notes on the hercules beetle, *Dynastes hercules* (Linn.), at Rancho Grande, Venezuela, with special references to combat behavior. *Zoologica,* 32: 09–116 + 4 plates.

Bristowe, W. S. 1976. *The World of Spiders.* London: Collins.

Craig, G. B., Jr. 1967. Mosquitoes: female monogamy induced by male accessory gland substance. *Science,* 156:1499–1501.

Eberhard, W. G. 1980. Horned beetles. *Scientific American,* 242:166–182.

———— 1985. *Sexual Selection and Animal Genitalia.* Cambridge, Mass.: Harvard University Press.

Happ, G. M. 1969. Multiple sex pheromones of the mealworm beetle, *Tenebrio molitor* L. *Nature,* 222: 80–181.

Knab, F. 1906. The swarming of *Culex pipiens. Psyche,* 13:123–133.

Lack, D. L. 1943. *The Life of the Robin.* London: H. F. and G. Witherby.

Mellanby, K. 1939. Fertilization and egg production in the bed-bug, *Cimex lectularius. Parasitology,* 31:193–199.

Metcalf, R. L., and R. A. Metcalf. 1994. Attractants, repellents, and genetic control in pest management. In R. L. Metcalf and W. H. Luckmann, *Introduction to Insect Pest Management*, 3rd ed. New York: John Wiley and Sons.

Smith, R. L., ed. 1984. *Sperm Competition and the Evolution of Animal Mating Systems*. New York: Academic Press.

Sturm, H. 1956. Die Paarung beim Silberfischchen (Pairing by silverfish). *Zeitschrif für Tierpsychologie*, 13:1–12.

Teale, E. W. 1949. *The Insect World of J. Henri Fabre*. New York: Dodd, Mead.

Thornhill, R. 1983. *The Evolution of Insect Mating Systems*. Cambridge, Mass.: Harvard University Press.

Waage, J. K. 1979. Dual function of the damselfly penis: sperm removal and transfer. *Science*, 203:916–918.

Wallace, A. R. 1869. *The Malay Archipelago*. Reprinted by Dover Publications, New York, in 1962.

Caring for Offspring

Borkhausen, M. B. 1790. *Systematische Beschreibung der europäischen Schmetterlinge* (*Systematic Description of European Moths and Butterflies*). Vol. 3. Frankfurt.

Buxton, P. A. 1955. *The Natural History of Tse Tse Flies*. London: Lewis and Co.

Cambefort, Y. 1987. Le scarabée dans l'Égypte ancienne (The scarab in ancient Egypt). *Revue de l'Histoire des Religions*, 204:3–46.

Cherry, R. H. 1985. Insects as sacred symbols in ancient Egypt. *Bulletin of the Entomological Society of America*, 31:14–16.

Cooper, K. W. 1954. Biology of eumenine wasps, IV. A trigonalid wasp parasitic on *Rygchium rugosum* (Saussure). *Proceedings of the Entomological Society of Washington*, 56:280–288.

Dambach, C. A., and E. Good. 1943. Life history and habits of the cicada killer in Ohio. *The Ohio Journal of Science*, 43:32–41.

Denlinger, D. L., and J. Zdárek. 1992. Rhythmic pulses and haemolymph pressure associated with parturition and ovulation in the tsetse fly, *Glossina morsitans*. *Physiological Entomology*, 17:127–130.

Dunn, L. H. 1930. Rearing the larvae of *Dermatobia hominis* Linn., in man. *Psyche*, 37:327–342.

Fink, D. E. 1915. *The Eggplant Lace-Bug*. Bulletin of the U.S. Department of Agriculture no. 239. Washington, D.C.

Frisch, K. von. 1953. *The Dancing Bees*. Translated from the German by Dora Ilse. New York: Harcourt, Brace and World.

Haldane, J. B. S. 1927. *Possible Worlds and Other Essays*. London: Chatto and Windus.

Hanski, I., and Y. Cambefort, eds. 1991. *Dung Beetle Ecology*. Princeton: Princeton University Press.

Hughes, L., and M. Westoby. 1992. Capitula on stick insect eggs and elaiosomes on seeds: convergent adaptations for burial by ants. *Functional Ecology*, 6:642–648.

Lenoble, B. J., and D. L. Denlinger. 1982. The milk gland of the sheep ked, *Melophagus ovinus*: a comparison with *Glossina*. *Journal of Insect Physiology*, 28:165–172.

Michelsen, A., B. B. Andersen, W. H. Kirchner, and M. Lindauer. 1989. Honeybees can be recruited by a mechanical model of a dancing bee. *Naturwissenschaften*, 76:277–280.

Morse, R., and T. Hooper. 1985. *The Illustrated Encyclopedia of Beekeeping*. New York: E. P. Dutton.

Scarbrough, A. G., G. P. Waldbauer, and J. G. Sternburg. 1974. Feeding and survival of cecropia (Saturniidae) larvae on various plant species. *Journal of the Lepidopterists' Society*, 28:212–219.

Seeley, T. D. 1985. *Honey Bee Ecology*. Princeton: Princeton University Press.

Tinbergen, H. K. 1972. *The Animal in Its World*. 2 vols. Cambridge, Mass.: Harvard University Press.

Tinbergen, N. 1951. *The Study of Instinct*. Oxford: Clarendon Press.

West, M. J., and R. D. Alexander. 1963. Sub-social behavior in a burrowing cricket, *Anurogryllus muticus* (DeGeer) (Orthoptera: Gryllidae.) *The Ohio Journal of Science*, 63: 9–24.

Wilson, E. O. 1971. *The Insect Societies*. Cambridge, Mass.: The Belknap Press of Harvard University Press.

Defense against Predators

Annandale, N. 1900. Observations on the habits and natural surroundings of insects made during the "Skeat" expedition to the Malay Peninsula. *Proceedings of the Zoological Society of London*, 1900:837–869.

Bates, H. W. 1862. Contributions to an insect fauna of the Amazon Valley, Lepidoptera: Heliconidae. *Transactions of the Linnaean Society, Zoology*, 23:95–566.

Blanchan, N. 1903. *Bird Neighbors*. New York: Doubleday, Page.

Blest, A. D. 1957. The function of eyespot patterns in the Lepidoptera. *Behaviour*, 11:209–256.

Breed, M. D., G. E. Robinson, and R. E. Page, Jr. 1990. Division of labor during honey bee colony defense. *Behavioral Ecology and Sociobiology*, 27:395–401.

Bristowe, W. S. 1958. *The World of Spiders*. London: Collins.

Brower, L. P. 1969. Ecological chemistry. *Scientific American*, 220:22–30.

Cott, H. B. 1957. *Adaptive Coloration in Animals*. London: Methuen.

Curio, E. 1966. Wie Insketen ihre Feinde abwehren (How insects ward off their enemies). *Naturwissenschaft und Medizin*, 11:3–22.

———— 1970. Validity of the selective coefficient of a behaviour trait in hawkmoth larvae. *Nature*, 222:382.

Edmunds, M. 1974. *Defence in Animals*. Essex, Eng.: Longman Group.

Eisner, T. 1972. Chemical ecology: on arthropods and how they live as chemists. *Verhandlungsbericht der Deutschen Zoologischen Gesellschaft*, 65. *Jahresversammlung*, 123:137.

Evans, D. L., and G. P. Waldbauer. 1982. Behavior of adult and naive birds when presented with a bumblebee and its mimic. *Zeitschrift für Tierpsychologie*, 59:247–259.

Fenton, M. B., and J. H. Fullard. 1981. Moth hearing and feeding strategies of bats. *American Scientist*, 69:266–275.

Hingston, R. W. G. 1932. *A Naturalist in the Guiana Forest*. New York: Longmans, Green.

Jacobson, E. 1911. Biological notes on the hemipteran *Ptilocerus ochraceus*. *Tijdschrift voor Entomologie*, 54:175–179.

Kettlewell, H. B. D. 1959. Darwin's missing evidence. *Scientific American*, 200:48–53.

Müller, F. 1879. *Ituna* and *Thyridis*; a remarkable case of mimicry in butterflies. Translated from the German by R. Meldola. *Proceedings of the Entomological Society of London*, 27:xx–xxix.

Newnham, A. 1924. The detailed resemblance of an Indian lepidopterous larva to the excrement of a bird. A similar result obtained in an entirely different way by a Malayan spider. *Transactions of the Entomological Society of London*, 1924:xc–xciv.

Owen, Denis. 1980. *Camouflage and Mimicry*. Chicago: University of Chicago Press.

Pough, F. H. 1988. Mimicry of vertebrates: are the rules different? In L. P. Brower, ed., *Mimicry and the Evolutionary Process*. Chicago: University of Chicago Press.

Sisson, R. F. 1980. Deception: formula for survival. *National Geographic*, 153:394–415.

Sternburg, J. G., G. P. Waldbauer, and M. R. Jeffords. 1977. Batesian mimicry: selective advantage of color pattern. *Science*, 195:681–683.

Stowe, M. K., J. H. Tumlinson, and R. R. Heath. 1987. Chemical mimicry: bolas spiders emit components of moth prey species sex pheromones. *Science*, 236:964–967.

Treat, A. E. 1975. *Mites of Moths and Butterflies*. Ithaca: Cornell University Press.

Waldbauer, G. P. 1988. Aposematism and Batesian mimicry. In M. K.Hecht, B. Wallace, and G. T. Prance, eds., *Evolutionary Biology*, 22:227–259.

———— 1988. Asynchrony between Batesian mimics and their models. In L.P. Brower, ed., *Mimicry and the Evolutionary Process*. Chicago: University of Chicago Press.

Whitman, D. W., M. S. Blum, and D. W. Alsop. 1990. Allomones: chemicals for defense. In D. L.Evans and J. O. Schmidt, eds., *Insect Defenses*. Albany: State University of New York Press.

Wickler, W. 1968. *Mimicry in Plants and Animals*. Translated from the German by R. D. Martin. New York: McGraw-Hill.

Winston, M. L. 1992. *Killer Bees: The Africanized Honey Bee in the Americas*. Cambridge, Mass.: Harvard University Press.

The Parasitic Way of Life

Askew, R. R. 1971. *Parasitic Insects*. New York: American Elsevier.

Cooper, K. W. 1954. Biology of eumenine wasps, IV. A trigonalid wasp parasitic on *Rygchium rugosum* (Saussure). *Proceedings of the Entomological Society of Washington*, 56: 80–288.

DeBach, P. 1974. *Biological Control by Natural Enemies*. Cambridge, Eng. Cambridge University Press.

Eibl-Eibesfeldt, I., and E. Eibel-Eibesfeldt. 1968. The workers' bodyguard. *Animals*, 11:16–17.

Farquharson, C. O. 1918. *Harpagomyia* and other Diptera fed by *Crematogaster* ants in S. Nigeria. *Transactions of the Entomological Society of London* 1918:xxix–xxxix.

Hassel, M. P. 1968. The behavioural response of a tachinid fly (*Cyzenis albicans* (Fall.)) to its host, the winter moth (*Operophtera brumata* (L.)). *Journal of Animal Ecology*, 37:627–639.

Hölldobler, B. 1976. Tournaments and slavery in a desert ant. *Science*, 192:912–914. The article is copyright 1976 by the AAAS.

Hölldobler, B. and E. O. Wilson. 1990. *The Ants*. Cambridge, Mass.: The Belknap Press of Harvard University Press.

Hölldobler, K. 1948. Über ein parasitologisches Problem. Die Gastpflege der Ameisen und die Symphilieinstinkte (On a parasitological problem. Guest care by ants and symphile instincts). *Zeitschrift für Parasitenkunde*, 14:3–26.

Huang, H. T., and Pei Yang. 1987. The ancient cultured citrus ant. *BioScience*, 37:665–671.

Marsh, F. L. 1937. Ecological observations upon the enemies of cecropia, with particular reference to its hymenopterous parasites. *Ecology*, 18:106–112.

Maschwitz, U., M. Wüst, and K. Schurian. 1975. Blauingsraupen als Zuckerliefernten für Ameisen (Lycaenid caterpillars as sugar sources for ants). With an English summary. *Oecologia* (Berlin), 18:17–21.

Metcalf, R. L., and R. A. Metcalf. 1993. *Destructive and Useful Insects*. 5th ed. New York: McGraw-Hill.

Monteith, L. G. 1955. Host preferences of *Drino bohemica* Mesh. (Diptera: Tachinidae), with particular reference to olfactory responses. *The Canadian Entomologist*, 87:509–530.

Newcomer, E. J. 1912. Some observations on the relations of ants and lycaenid caterpillars, and a description of the relational organs of the latter. *Journal of the New York Entomological Society*, 20:31–36 + 2 plates.

Pierce, N. E., and P. S. Mead. 1981. Parasitoids as selective agents in the symbiosis between lycaenid butterfly larvae and ants. *Science*, 211:1185–1187.

Redborg, K. E., and E. G. MacLeod. 1985. *The Developmental Ecology of "Mantispa uhleri" Banks (Neuroptera: Mantispidae)*. Illinois Biological Monographs. Urbana: University of Illinois.

Ristich, S. S. 1953. A study of the prey, enemies, and habits of the great-golden digger wasp *Chlorion ichneumoneum*. *The Canadian Entomologist*, 85:374–386.

Salt, G. 1968. The resistance of insect parasitoids to the defence reaction of their hosts. *Biological Review*, 43:200–232.

Sternburg, J. G., G. P. Waldbauer, and A. G. Scarbrough. 1981. Distribution of the cecropia moth (Saturniidae) in central Illinois: a study in urban ecology. *Journal of the Lepidopterists' Society*, 35:304–320.

Thornhill, R. 1975. Scorpionflies as kleptoparasites of web-building spiders. *Nature*, 258:709–711.

Turlings, T. C. J., J. H. Tumlinson, and W. J. Lewis. 1990. Exploitation of herbivore-induced plant odors by host-seeking parasitic wasps. *Science*, 250:1251–1253.

Waage, J. K., and G. G. Montgomery. 1976. *Cryptoses choloepi:* a coprophagous moth that lives on a sloth. *Science,* 193:157–158.

Wheeler, W. M. 1910. *Ants, Their Structure, Development, and Behavior.* New York: Columbia University Press.

Vollrath, F. 1979. Behaviour of the kleptoparasitic spider *Argyodes elevatus* (Araneae, Theridiidae). *Animal Behavior,* 27:515–521.

Recognizing Food

Abrahamson, W. G., ed., 1989. *Plant-Animal Interactions.* New York: McGraw-Hill.

Aker, C. L. and D. Udovic. 1981. Oviposition and pollination behavior of the yucca moth, *Tegeticula maculata* (Lepidoptera: Prodoxidae), and its relation to the reproductive biology of *Yucca whipplei* (Agavaceae). *Oecologia* (Berlin), 49:96–101.

Bates, M. 1949. *The Natural History of Mosquitoes.* New York: MacMillan.

Belt, T. 1888. *The Naturalist in Nicaragua.* London: Edward Bumpus.

Berenbaum, M. R., and E. Miliczky. 1984. Mantids and milkweed bugs: efficacy of aposematic coloration against invertebrate predators. *American Midland Naturalist,* 111:64–68.

Clark, L., and J. R. Mason. 1988. Effect of biologically active plants used as nest material and the derived benefit to starling nestlings. *Oecologia* (Berlin), 77:174–180.

Forde, D., ed. 1956. *Efik Traders of Old Calabar.* London: Oxford University Press.

Fraser, T. R. 1864. On the moth of the esere, or ordeal-bean of old Calabar. *Annals and Magazine of Natural History,* 13:389–393.

Futuyuma, D. J., and M. Slatkin, eds. 1983. *Coevolution.* Sunderland, Mass.: Sinauer Associates.

Janzen, D. H. 1966. Coevolution of mutualism between ants and acacias in Central America. *Evolution,* 20:249–275.

Lewis, W. H., and M. P. F. Elvin-Lewis. 1977. *Medical Botany.* New York: John Wiley and Sons.

Metcalf, R. L., and R. A. Metcalf. 1993. *Destructive and Useful Insects.* New York: McGraw-Hill.

Nicholson, R. 1992. Death and Taxus. *Natural History,* 101:20–23.

Rilling, S., H. Mittelstaedt, and K. D. Roeder. 1959. Prey recognition in the praying mantis. *Behaviour,* 14:164–184.

Ruiter, L. de. 1952. Some experiments on the camouflage of stick caterpillars. *Behaviour,* 4:222–232.

Swain, T., ed. 1972. *Plants in the Development of Modern Medicine.*
Cambridge, Mass.: Harvard University Press.
Thorsteinson, A. J. 1953. The chemotactic responses that determine host
specificity in an oligophagous insect (*Plutella maculipennis* (Curt.)
Lepidoptera). *Canadian Journal of Zoology,* 31:52–72.
Tinbergen, N. 1951. *The Study of Instinct.* Oxford: Clarendon Press.
Wasson, R. G. 1968. *Soma, Divine Mushroom of Immortality.* New York:
Harcourt, Brace and World.
Witt, P. N., and C. F. Reed. 1965. Spider-web building. *Science,*
149:1190–1197.

Taking Nourishment

Baer, W. S. 1931. The treatment of chronic osteomyelitis with the maggot
(larva of the blowfly). *Journal of Bone and Joint Surgery,* 13:438–475.
Bänziger, H. 1971. Bloodsucking moths of Malaya. *Faunna,* 1:4–16.
——— 1980. Skin-piercing blood-sucking moths III: feeding act and
piercing mechanism of *Calyptra eustrigata* (Hmps.) (Lep.,
Noctuidae). *Bulletin de la Société Entomologique Suisse,* 53:127–142.
Barer-Stein, T. 1980. *You Eat What You Are.* Toronto: McClelland and
Stewart.
Bodenheimer, R. S. 1951. *Insects as Human Food.* The Hague: W. Junk.
Kennedy, J. S., and T. E. Mittler. 1953. A method of obtaining phloem sap
via the mouth-parts of aphids. *Nature,* 171:528.
Lindlahr, V. H. 1940. *You Are What You Eat.* New York: National Nutrition
Society.
Montgomery, S. L. 1982. Biogeography of the moth genus *Eupithecia* in
Oceania and the evolution of ambush predation in Hawaiian
caterpillars (Lepidoptera: Geometridae). *Entomologia Generalis,*
8:27–34.
Oldroyd, H. 1964. *The Natural History of Flies.* New York: W. W. Norton.
Robinson, W. 1935. Allantoin, a constituent of maggot excretions,
stimulates healing of chronic discharging wounds. *Journal of
Parasitology,* 21:354–358.
Rothschild, M., and T. Clay. 1952. *Fleas, Flukes, and Cuckoos.* London:
Collins.
Sisson, R. F. 1980. Deception: formula for survival. *National Geographic,*
153:394–415.
Usinger, R. L. 1966. *Monograph of the Cimicidae* (Hemiptera-Heteroptera).
College Park, Md.: Entomological Society of America.

Wigglesworth, V. B. 1972. *The Principles of Insect Physiology.* 7th ed. London: Chapman and Hall.

Zinsser, H. 1935. *Rats, Lice, and History.* Boston: Little, Brown.

Coping with the Seasons

Beck, S. D. 1980. *Insect Photoperiodism.* 2nd ed. New York: Academic Press.

Dingle, H. 1972. Migration strategies of insects. *Science,* 175:1327–1335.

Dingle, H., ed., 1978. *Evolution of Insect Migration and Diapause.* New York: Springer-Verlag.

Krysan, J. L., D. E. Foster, T. F. Branson, K. R. Ostlie, and W. S. Cranshaw. 1986. Two years before the hatch: rootworms adapt to crop rotation. *Bulletin of the Entomological Society of America,* 32:250–253.

Porter, G. S. 1909. *A Girl of the Limberlost.* New York: Grosset and Dunlap.

———— 1912. *Moths of the Limberlost.* Garden City, N.Y.: Doubleday, Page.

Riley, J. R., X-N. Cheng, X-X. Zhang, D. R. Reynolds, G-M. Xu, A. D. Smith, J-Y. Cheng, A-D. Bao, and B-P. Zhai. 1991. The long distance migration of *Nilaparvata lugens* (Stål) (Delphacidae) in China: Radar observations of mass return flight in the autumn. *Ecological Entomology,* 16:471–489.

Salt, R. W. 1961. Principles of insect cold-hardiness. *Annual Review of Entomology,* 6:55–74.

Tuskes, P. M., and L. P. Brower. 1978. Overwintering ecology of the monarch butterfly, *Danaus plexippus* L., in California. *Ecological Entomology,* 3:141–153.

Urquhart, F. A. 1976. Found at last: the monarch's winter home. *National Geographic,* 150:160–173.

Waldbauer, G. P., and J. G. Sternburg. 1973. Polymorphic termination of diapause by cecropia: genetic and geographical aspects. *Biological Bulletin,* 145:627–641.

Wigglesworth, V. B. 1972. *The Principles of Insect Physiology.* London: Chapman and Hall.

Zahl, P. A. 1963. Mystery of the monarch butterfly. *National Geographic,* 123:588–598.

Silken Cocoons

Feltwell, John. 1990. *The Story of Silk.* New York: St. Martin's Press.

Hiratsuka, E. 1920. Researches on the nutrition of the silkworm. *Bulletin of the Imperial Sericultural Experiment Station, Japan,* 1:257–315.

Hölldobler, B., and E. O. Wilson. 1990. *The Ants.* Cambridge, Mass.: The Belknap Press of Harvard University Press.

Hook, B. 1982. *The Cambridge Encyclopedia of China.* Cambridge, Eng.: Cambridge University Press.

Porter, G. S. 1912. *Moths of the Limberlost.* Garden City, N.Y.: Doubleday, Page.

Scarbrough, A. G., J. G. Sternburg, and G. P. Waldbauer. 1977. Selection of the cocoon spinning site by the larvae of *Hyalophora cecropia* (Saturniidae). *Journal of the Lepidopterists' Society,* 31:153–166.

Scarbrough, A. G., G. P. Waldbauer, and J. G. Sternburg. 1972. Response to cecropia cocoons of *Mus musculus* and two species of *Peromyscus.* *Oecologia* (Berlin), 10:137–144.

Van der Kloot, W. G. 1956. Brains and cocoons. *Scientific American,* 194:131–140.

Waldbauer, G. P., A. G. Scarbrough, and J. G. Sternburg. 1982. The allocation of silk in the compact and baggy cocoons of *Hyalophora cecropia. Entomologia Experimentalis et Applicata,* 31:191–196.

Waldbauer, G. P., and J. G. Sternburg. 1982. Cocoons of *Callosamia promethea* (Saturniidae): adaptive significance of differences in mode of attachment to the host tree. *Journal of the Lepidopterists' Society,* 36: 92–199.

Waldbauer, G. P., J. G. Sternburg, W. G. George, and A. G. Scarbrough. 1970. Hairy and downy woodpecker attacks on cocoons of urban *Hyalophora cecropia* and other saturniids (Lepidoptera). *Annals of the Entomological Society of America,* 63:1366–1369.

Winter

Porter, G. S. 1912. *Moths of the Limberlost.* Garden City, N.Y.: Doubleday, Page.

Scarbrough, A. G., G. P. Waldbauer, and J. G. Sternburg. 1972. Response to cecropia cocoons of *Mus Musculus* and two species of *Peromyscus.* *Oecologia* (Berlin), 10:137–144.

Sternburg, J. G., G. P. Waldbauer, and J. G. Scarbrough. 1981. Distribution of cecropia moths (Saturniidae) in central Illinois: a study in urban ecology. *Journal of the Lepidopterists' Society,* 35:304–320.

Acknowledgments

I am indebted to the many friends and colleagues who helped me, giving generously of their time and knowledge: May Berenbaum, Jeanine Berlocher, Murray Blum, John Bouseman, Jeff Brawn, Lincoln Brower, George Craig, David Denlinger, Hugh Dingle, Susan Fahrbach, Stanley Friedman, William Horsfall, Robert Lewis, James Lloyd, Ellis MacLeod, Steve Malcolm, Robert Novak, John Pinto, Hugh Robertson, Gene Robinson, Felipe Soto-Adames, James Sternburg, Ro Vaccaro, and Judy Willis.

   Julia Berger, Phyllis Cooper, Annie Laurie Horsfall, Ellis MacLeod, James Nardi, and James Sternburg provided constructive criticism of parts of the manuscript, and May Berenbaum and my wife, Stephanie Waldbauer, critically read it in its entirety. Dorothy Nadarski patiently typed and retyped the manuscript. Amy Bartlett Wright was meticulous in her execution of the drawings. The book benefited greatly from the constant encouragement of Michael Fisher and the careful editing of Nancy Clemente.

Index

Page numbers in **bold** refer to figures.

Library of Congress Cataloging-in-Publication Data

Waldbauer, Gilbert.
    Insects through the Seasons / Gilbert Waldbauer.
       p.   cm.
    Includes bibliographical references (p.   ) and index.
    ISBN 0-674-45488-X (hardcover : alk. paper)
    1. Insects—Behavior.   2. Cecropia moth—Behavior.   3. Insects—Life cycles.
    4. Cecropia moth—Life cycles.   5. Seasons.   I. Title.
  QL496.W34   1996
  595.7′051—dc20   95-35171